항공영어

저자 조용욱
한석태

도서출판 청연

머 리 말

　첨단 기술의 집합체라 일컬어지는 항공 산업은 그 나라의 과학 기술 수준의 척도라 해도 과언이 아니다. 다행히도 우리나라도 근래에 들어 적극적인 항공 기술 사업을 추진하기 시작했다. 여타 첨단 기술과 마찬가지로 항공 기술 역시 세계에서 몇 안되는 나라에서 도입되었으며, 앞으로도 계속될 것으로 믿고 있다.

　우리나라 사람들 모두가 인지하다시피 외국어는 세계화 추세에 맞추어 필수 도구가 되고 있다. 굳이 거창한 수사는 차치하더라도 현실적으로 우리의 독자 여러분은 항공기의 기술적인 면에 관심이 있고 배우고자 하는 것이기 때문에 필요 불가결하게 원문을 접하게 될 것이다. 현재 우리나라에는 외국어 교육 자료가 수없이 많이 있으나, 항공 기술에 관한 것은 거의 없는 실정이다. 물론 기술 영어도 일반 교양 영어와 그 기본틀에서 큰 차이점은 없으나, 영문학을 전공한 학생이 항공 기술 분야의 외국 자료를 제대로 이해하지 못하는 것은 바로 기술 영어의 특성이 있기 때문이다.

　이에 저자는 이제 막 항공기에 대해 공부하고자 하는 학도들에게 항공 기술에 대한 이해를 돕고자 본 항공영어 책자를 발간하게 되었다. 여러가지 미비한 점이 많이 있을 줄 알고 있으나 제한된 분량에 모든 항공 분야에 대한 것을 다루지는 못했다.

　따라서 책의 구성은 항공 기술 전문교육기관에서 배우기에 적합하도록 편집하려고 힘썼다. 제1, 2, 3장은 간단한 영어 원문을 제시하고 일부 단어의 해설과 본문 해석에 중점을 두어 서서히 친숙해지도록 하였고, 제4장은 상용 항공기의 각 시스템 설명을 원문대로 실어서 스스로 연구할 수 있도록 하였다. 첨언하자면 원문에 나오는 단어는 항공 기술 영어의 뜻을 풀이해놓았으므로, 그 외 일반 영어로서의 여러가지 뜻이 있음을 주지하고 폭넓은 의미를 스스로 탐구하길 바란다. 부디 학도 여러분의 노력에 응당한 결실이 맺길 기대한다.

1994. 5. 15.

저　　자

목 차

제1장 AIRCRAFT GENERAL

1-1. General ──────────────────────────── 1
  1) Aircraft and Airplane ──────────────── 1
  2) Fuselage ─────────────────────── 2
  3) Wing ────────────────────────── 6
  4) Flight Control Surface ──────────────── 8

1-2. Helicopter ─────────────────────────── 16
  1) General ──────────────────────── 16
  2) Principles of Flight ────────────────── 19
  3) Helicopter Components and Function ─────── 21

1-3. Jet Engine ─────────────────────────── 25
  1) Jet Propulsion ───────────────────── 25
  2) Engine Thrust Setting ──────────────── 31
  3) Fuel Consumption ─────────────────── 33

제2장 AIRCRAFT OPERATION

2-1. Cockpit Operations ───────────────────── 35
  1) Starting Engine ──────────────────── 35
  2) Taxiing ───────────────────────── 36
  3) Before Takeoff ──────────────────── 36
  4) After Takeoff ───────────────────── 37
  5) Climb ────────────────────────── 38
  6) Cruise ───────────────────────── 39
  7) Approach ─────────────────────── 40
  8) Before Landing ──────────────────── 40
  9) Landing ──────────────────────── 41

2-2. Radio Communication ──────────────── 43
   1) Rules for Radio Communications ─────────── 43
   2) Radiotelephony ───────────────────── 46

2-3. Passenger Service ─────────────────── 55
   1) Information on Baggage ──────────────── 55
   2) Passenger Address in Cabin ──────────── 56

2-4. Aircraft Accident and Hijacking ─────────── 61
   1) Aircraft Accidents ───────────────── 61
   2) Hijacking ──────────────────────── 64
   3) Emergancy Procedures ──────────────── 66

제3장 AIRCRAFT MAINTENANCE

3-1. Maintenance Manual ─────────────────── 71
   1) Organization of Maintenance Manual ─────── 71
   2) Definitions of Maintenance Terms ────────── 73
   3) Purpose of Aircraft Service System ──────── 78
   4) Minimum Equipment Requirement ────────── 79

3-2. Preflight Check and Service Work Items ─────── 81
   1) Cockpit and Cabin ──────────────────── 81
   2) Aircraft Exterior and Interior General ──────── 83
   3) Engine, Pod and Pylon ──────────────── 84

3-3. Service Work Items ─────────────────── 86
   1) Inspection Skill ───────────────────── 86
   2) Aircraft Skill ────────────────────── 94
   3) Hydraulic Skill ───────────────────── 95
   4) Engine Skill ─────────────────────── 98
   5) Electrical Skill ──────────────────── 100

3-4. Discrepancy ---105
   1) Ground Discrepancies ----------------------------------- 105
   2) Flight Discrepancies ----------------------------------- 115

3-5. Modification or Alteration -----------------------------117
   1) Service Bulletin --------------------------------------- 117
   2) Airworthiness Directives ------------------------------- 118

3-6. Inspection System of FAA Repair Station --------122
   1) General Manager Inspection Department ------------------ 122
   2) Inspection Personnel ----------------------------------- 126
   3) Record of Work and Inspection -------------------------- 127
   4) Control of Precision Measurement Equipment and
      Measurements Standard ----------------------------------128
   5) Malfunction or Defect Report --------------------------- 129
   6) Type of Inspection ------------------------------------- 129
   7) Inspection of Critical Items --------------------------- 131
   8) Non-Destructive Inspection(NDI) ------------------------ 132
   9) Handling of Parts -------------------------------------- 133
   10) Tagging and Identification of Parts ------------------- 134
   11) Inspection Stamps ------------------------------------- 136

3-7. Ground Support Equipment ------------------------------138
   1) G.S.E. for Maintenance --------------------------------- 138
   2) Stands and Ladders for Maintenance --------------------- 141

제4장 AIRCRAFT SYSTEM DESCRIPTION

4-1. Types, Design Features and Configurations of
   Transport Aircraft --------------------------------------145
   1) Boeing 737 Series -------------------------------------- 145
   2) Boeing 747 Series -------------------------------------- 148
   3) 767 Twin Jet --- 153

4) McDonnell Douglas MD-80 Series ──────────── 156
5) Douglas DC-10 Series ──────────────────── 159
6) McDonnell Douglas MD-11 ──────────────── 162
7) Airbus A320 ────────────────────────── 165

4-2. Auxiliary Power Units, Pneumatic, and Environmental Control Systems ──────174
1) APU Systems ─────────────────────────── 174
2) Pneumatic Systems ───────────────────── 176
3) Environment Control Systems ──────────── 177
4) Pressurization Systems ──────────────── 180

4-3. Anti-icing Systems and Rain Protection ────────189
1) Aircraft Ground Deicing/Anti-icing ─────── 190
2) Boeing 757 Aircraft Anti-icing Systems ──── 193

4-4. Electrical Power Systems ─────────────────198
1) Power Sources ──────────────────────── 198
2) System Components ──────────────────── 199
3) Electrical System Configurations ────────── 200
4) Boeing 747 Electrical Power System ──────── 203

4-5. Flight Control Systems ────────────────── 209
1) Boeing 727 ─────────────────────────── 209

4-6. Fuel Systems ──────────────────────────218
1) Turbine Engine Fuels ───────────────────── 218
2) Fuel System Contamination ────────────── 218
3) Fuel Systems ────────────────────────── 220
4) Fuel System components and Subsystems ──── 220
5) Boeing 747-400 Fuel System ─────────────── 227

4-7. Hydraulic Systems ────────────────────── 234
1) Hydraulic Fluid ─────────────────────── 234

2) Components of Hydraulic Systems ------------------------- 235
3) Boeing 757 Hydraulic System ---------------------------- 240
4) Boeing 757 Landing Gear System ------------------------ 247

4-8. Oxygen Systems --259
1) Lockheed L-1011 Oxygen System ------------------------ 261

4-9. Warning and Fire Protection System ------------269
1) Fire Protection Systems -------------------------------- 269
2) Lockheed L-1011 Fire Pretection Systems --------------- 275
3) Aircraft Warning Systems ------------------------------ 283
4) L-1011 Aural Warning Systems ------------------------- 283
5) Boeing 737-300 Fire and Warning Systems -------------- 283

4-10. Communications, Instruments, and Navigational Systems --299
1) Communications --------------------------------------- 299
2) Navigation Equipment --------------------------------- 304
3) Boeing 757 Avionics Systems --------------------------- 307

4-11. Miscellaneous Aircraft Systems and Maintenance Information --326
1) Portable Water Systems -------------------------------- 326
2) Waste Systems -- 330
3) Lighting Systems -------------------------------------- 335
4) Emergency Equipment --------------------------------- 340
5) Equipment Cooling Systems ---------------------------- 346

제1장 AIRCRAFT GENERAL

항공에 관한 영문을 읽으려면 무엇보다도 항공기란 어떤 것이고 어떤 기능을 가지고 있는 것인가를 먼저 알아둘 필요가 있다. 따라서 여기서는 항공기의 개요를 이해해 보기로 한다.

1-1. General

1) Aircraft and Airplane

> **1** Aircraft : In a broad sense, any machine or craft designed to go through the air(including, in some instances, outer space), given lift by its own buoyancy(as with airships), or by dynamic reaction of air particles over and about its surfaces, or by reaction of a jet stream or other fluid jet.

〈단어 연구〉
broad 광범위한, 광대한
sense 관념, 의미
in a broad sense 넓은 의미로는
craft 비행기
in some instances 몇가지 예 중에는
including은 outer space와 계속(외계의 우주를 포함)된다.
lift 양력
buoyancy 부력
given lift by ~는 machine or craft에 걸림
as with airships 비행선이 갖고 있는 것과 같이
dynamic 동적, 역학상의
reaction 반동, 반작용
particle 분자, 부스러기
jet 젯트
stream 흐름
fluid 유동성의, 기체

제1장 Aircraft General

〈해 석〉
항공기 : 넓은 의미로는(비행선이 가지고 있는 것과 같이) 자기 자신의 부력이나 그 상면 또는 주변에서의 공기 분자의 역학적 반작용 혹은 젯트 분출 흐름이라든가 다른 유체의 분출에 따른 반동에 의해 양력을 만들어 공중(몇가지 예 중에서는 외계의 우주를 포함하고 있음)을 항행하도록 설계된 기계 또는 비행체를 말한다.

> **2** Orville and Wilbur Wright were the first to build and the first to fly in a powered heavier-than-air machine. This important event in aviation history took place on Dec. 17, 1903, at Kitty Hawk, North Carolina.

〈단어 연구〉
heavier-than-air 공기보다 무거운
event 대사건
take place 일어나다(happen, come to pass, occur, when do you expect it to take place?)

〈해 석〉
오빌과 윌버 라이트는 동력 구동의 공기보다 무거운 기계를 조립하여 그것을 타고 비행한 최초의 사람이다. 항공 역사상 이 중요한 사건은 1903년 12월 17일에 노스 케롤라이나주의 키티 호크에서 일어났다.

2) Fuselage(기체)

> **1** The fuselage is the main structure or body of the aircraft. It provides space for cargo, controls, accessories, passengers, and other equipment.

〈단어 연구〉
body 본체, 동체
provide ~을 제공하다(to furnish, supply, or equip, to afford or yield, 가끔 뒤에 for가 붙는다. to provide for one's family)
space 스페이스, 구역, 공간
cargo 화물

controls 조종 장치
accessories (복수형) 부속품
passenger 승객
equipment 장비품

〈해 석〉
동체는 항공기의 주요 구조 또는 본체이다. 동체는 화물, 조종 장치, 부속품, 승객 그리고 기타 장비품에 공간을 제공한다.

> ② The strong, heavy longerons hold the bulkheads and formers, and these, in turn, hold the stringers. All of these joined together form a rigid fuselage framework.

〈단어 연구〉
strong 튼튼한
heavy 무거운
hold 유지하다(비슷한 말로는 persist, last, endure)
in turn 순서대로, 다음에는
form 형성하다
framework (전체의) 틀, 구조

〈해 석〉
강하고 무거운 론저론은 벌크헤드나 퍼머(성형 부재)를 지탱하고 이들(벌크헤드나 퍼머)이 또 스트링거를 지탱한다. 이들 모두가 결합되어 전체로서 견고한 동체 프레임 구조를 형성한다.

> ③ The metal skin or covering is riveted to longeron, bulkheads, and other structural members and carries part of load. The fuselage skin thickness will vary with the load carried and the stresses sustained at a particular location.

〈단어 연구〉
rivet 리벳으로 고정하다
part 일부분

4   제1장 Aircraft General

thickness 두께
vary (여러가지로) 변하다[to change or alter, as in form, appearance, character, or substanc, ex) opinions vary on the outcome]
sustain 견디다(to support, hold, or bear up from below, 비슷한 말로는 carry, bear, maintain 등이 있다)
particular 특정의

〈해 석〉
금속 외판 또는 덮개는 론저론, 벌크헤드 및 기타 구조 부재에 리벳으로 고정되어 하중의 일부분을 지탱한다. 동체 외판의 두께는 지탱해야 할 하중, 그리고 어떤 특정한 위치에서 견뎌야 할 응력에 따라 다르다.

4 There are various numbering system in use to facilitate location of specific wing frames, fuselage bulkheads, or any other structural members on an aircraft. A typical station diagram is shown in Fig. 1-1.

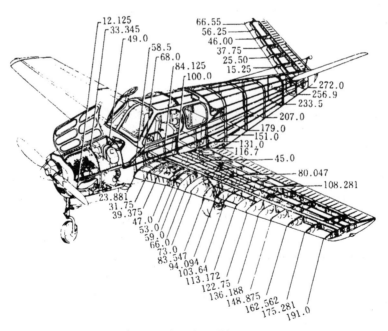

그림 1-1 Station Diagram

〈단어 연구〉
    number 번호를 붙이다
    system 본체
    facilitate (일을) 용이하게 하다(1.to make easier or less difficult. 2.to assist
      the progress of a person)
    location 위치 설정
    specific 특별한
    station diagram 스테이션 다이어그램, 위치 도표

〈해 석〉
    항공기에서는 특정한 날개 프레임, 동체의 벌크헤드 또는 기타 구조 부재 위치 설정을 편리하게 하기 위해 여러가지 번호 시스템이 사용되고 있다. 전형적인 위치 도표가 그림 1-1에 나타내어진다.

> **5** Fuselage stations(Fus. Sta. or F.S) are numbered in inches from a reference or zero point known as the reference datum. The reference datum is an imaginary vertical plane at or near the nose of the aircraft from which all horizontal distances are measured.

〈단어 연구〉
    fuselage station 동체 위치(Fus. Sta. 또는 F.S로 약함)
    reference or zero point 기준점 또는 영점
    known as ~로 알려져 있다
    reference datum 기준 위치(datum point에서 「기준점」으로서 통상 쓰이고 있다)
    imaginary 가공의
    vertical plane 수직면
    at or near the nose of the aircraft 항공기의 기수부 또는 그 부근에
    from which 가공의 수직면에서
    horizontal distance 수평 거리

〈해 석〉
    동체 위치는 기준 위치로 알려져 있는 기준점 또는 영점으로부터 인치로 번호가 붙여진다. 기준 위치는 항공기의 기수부 또는 그 부근에 있는데, 그곳에서부터 모든 수평 거리가 측정되는 가공의 수직면이다.

6   제1장 Aircraft General

3) Wing

> **1** Wing : The full cantilever wing is of all-metal stressed-skin construction. The primary structure of the inner wing panels is a box with two main beams. The upper surface is covered with aluminum alloy skin stiffened internally with corrugations, while the lower surface is a machined skin with integral stiffeners. Ribs are installed at frequent intervals to stabilize the structure and maintain contour.

〈단어 연구〉

```
cantilever  긴 봉이나 판을 한쪽 끝 만을 고정하여 전체를 지탱하고 있는 구조를 말함
stressed-skin  응력 외피, 하중의 일부를 담당하는 외피를 말함
construction  구조, 조립
beam  선복
aluminum  알루미늄
stiffen  강하게 하다
corrugation  파상, 파형
machined  기계 가공한
integral  완전한
stiffener  보강재, 스티프너
contour  윤곽, 등고
```

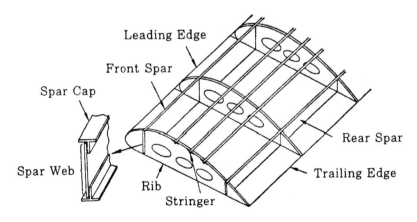

그림 1-2 Wing Structure

〈해 석〉

날개 : 완전한 외팔보식의 날개는 전금속제 응력 외피 구조이다. 안쪽 날개의 일차 구조는 2개의 메인 스파(spar)가 있는 상자형 구조이다. 상면은 내측에 파형재로 보강된 알루미늄 합금의 외판으로 커버되는 반면 밑면은 완전한 많은 스티프너와 하나로 되어 가공한 판이다. 리브는 구조를 안정시켜 그 윤곽을 유지하도록 짧은 간격으로 조립된다 (그림 1-2 참조).

---

2 An airfoil is technically defined as any surface, such as an airplane aileron, elevator, rudder, or wing, designed to obtain reaction from the air through which it moves. An airfoil section is a cross section of an airfoil.

〈단어 연구〉

airfoil 에어포일, 익형
aileron 에일러론, 보조 날개
elevator 엘레베이터, 승강타
rudder 러더, 방향타
reaction 반력(a reverse movement or tendency ; an action in a reverse direction or manner)

〈해 석〉

에어포일은 전문적으로는 공기중을 이동할 때 공기에서 반력을 얻을 수 있게 설계된 비행기의 조종 날개, 승강타, 방향타 또는 날개와 같은 면으로 정의된다. 익단면이란 날개의 절단면이다.

---

3 A flat plate perpendicular to an airstream obtains reactions from the air through which it moves, but it is not useful as an airfoil because there is no lift and turbulence is excessive. When an airfoil is at an acute angle(less than a right angle) to the airstream, it is an airfoil because there is lift.

〈단어 연구〉

excessive 과대한(exceeding the usual or proper limit or degree ;

characterized by excess, 비슷한 말로는 extravagant, extreme, inordinate, unreasonable 등이 있다)
an acute angle 예각
less than ~ 미만의
right angle 직각

〈해 석〉

기류와 직교하는 평판은 공기중을 이동할 때 공기에서 반력을 얻는데 양력을 일으키지 않고 흐트러짐이 심하므로 날개로서는 역할을 못한다. 날개는 기류에 대해 예각(직각 미만)에 있을 때 양력이 생기므로 에어포일이라고 한다.

4) Flight Control Surface

1️⃣ The flight control surfaces are hinged or movable airfoils designed to change the attitude of the aircraft during flight. These surfaces may be divided into three groups, usually referred to as the primary group, secondary group, and auxiliary group.

〈단어 연구〉

flight control surface 조종면
hinge 돌쩌귀를 달다
airfoil 에어포일, 익형
attitude 자세
divide (몇개의 부분으로) 나누다
referred to as ~라고 불리우다
primary 주요한
secondary 부의, 종의
auxiliary 보조의

〈해 석〉

조종면은 비행중 항공기의 자세를 바꿀 수 있도록 설계된 힌지가 달린 경우든지 혹은 가동형의 에어포일이다. 이들 조종면은 3개의 그룹으로 나눌 수 있는데, 일반적으로 1차 그룹, 2차 그룹 및 보조 그룹이라고 불리운다.

> **2** Primary group includes the ailerons, elevators, and rudder(Fig.1-3). These surfaces are used for moving the aircraft about its three axes. The ailerons and elevators are generally operated from the cockpit by a wheel and yoke assembly. The rudder is operated by foot pedals.

〈단어 연구〉
　three axes　(종, 횡, 정방향의) 3축
　axes　axis의 복수형
　operate　조종하다
　cockpit　조종실
　wheel　차륜
　yoke　조종간
　foot pedal　발로 밟는 페달(러더 페달)

〈해 석〉
　1차 그룹에는 보조 날개, 승강타 및 방향타가 있다. 이들 조종면(타면)은 항공기를 3축 주위로 움직이는데 사용된다. 보조 날개 및 승강타는 일반적으로 조종간에 의해 조종실에서 조작된다. 러더는 발로 밟는 페달로 조작된다.

> **3** Included in the secondary group are the trim tabs and spring tabs. Trim tabs are small airfoils recessed into the trailing edges of the primary control surfaces. The purpose of trim tabs is to enable the pilot to trim out any unbalanced condition which may exist during flight, without exerting any pressure on the primary controls.

〈단어 연구〉
　Included are~　도치문으로, 주어는 trim tabs and spring tabs이다
　trim tab　트림 탭(조종하여 비행할 수 있도록 조절하는 탭)
　spring tab　스프링 탭
　recess~　오목하게 패인 곳, 휴식
　purpose　용도, 목적
　is to~　하는데 있다, ~하기로 되어 있다.
　trim　(기체의) 균형을 유지하다

10　제1장　Aircraft General

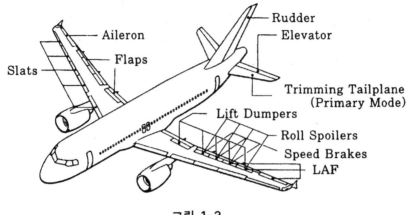

그림 1-3

```
out~    해버려, 충분히
unbalanced   균형이 잡히지 않는
exert pressure on   ~에 압력을 가하다
```

〈해 석〉
2차 그룹에 포함되는 것에는 트림 탭 및 스프링 탭이 있다. 트림 탭은 주조종면의 트레일링에이지에 끼워진 작은 날개이다. 트림 탭의 목적은 비행중에 일어날 수 있는 불균형 상태를 파일럿이 1차 조종 장치에 힘을 가하지 않고 균형을 이루게 하는 것이다.

❹ The auxiliary groups may be divided into two subgroups. Those whose primary purpose is lift augmenting and those whose primary purpose is lift decreasing. In the first group are the flaps, both trailing edge and leading edge (slats), and slots. The lift decreasing devices are speed brakes and spoilers.

〈단어 연구〉
```
whose  (관계대명사 which의 소유격) ~하는 (곳의)
lift augmenting   양력을 증가시키는
lift decreasing   양력을 감소시키는
trailing edge   (날개의) 후연
```

```
leading edge  (날개의) 전연
slot  슬롯(슬롯 방식의 플랩)
speed brake  스피드 브레이크
spoiler  스포일러
```

〈해 석〉

보조 그룹은 두개의 서브 그룹으로 나뉘어진다. 주된 목적이 양력 증가인 그룹과 주된 목적이 양력 감소인 그룹이다. 전자의 그룹은 플랩, 전연, 후연, 두개의 슬랫(리딩에이지 플랩을 슬랫이라고 부름), 그리고 슬롯이다. 양력 저감 장치는 스피드 브레이크와 스포일러이다.

> **5** All the control surfaces of a large turbojet aircraft are shown in Fig.1-4. As illustrated, each wing has two ailerons, one in the conventional position at the outboard trailing edge of the wing and another hinged to the trailing edge of the wing center section.

〈단어 연구〉

```
as illustrated  도해되어 있는 것처럼
conventional  일반적인(conforming or adhering to accepted standards, as of
    conduct or taste 비슷한 말로는 usual, habitual customary)
outboard  날개 끝에 가까운 쪽의(↔inboard)
```

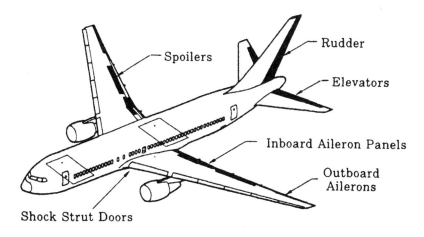

그림 1-4 Boeing 767

12 제1장 Aircraft General

〈해 석〉
대형 터보 젯트기의 조종면 전체를 그림 1-4에 나타낸다. 표시되어 있듯이 각 날개에는 두개의 보조 날개가 있는데, 하나는 날개 외측 트레일링에이지의 일반적인 위치에 그리고 다른 하나는 날개 중앙부 트레일링에이지에 힌지로 장착되어 있다.

> **6** Ailerons respond to side pressure applied to the control stick. Pressure applied to move the stick toward the right raises the right aileron and lowers the left aileron, causing the aircraft to bank to the right. Ailerons are linked together by control cables so that when one aileron is down, the opposite aileron is up.

〈단어 연구〉
respond to ~에 응답하다
side pressure 횡방향의 힘
control stick 조종간
raise (물건을) 들어올리다(to move to a higher position : lift 비슷한 말로는 lift, heave, hoist 등이 있다)
lower 낮게 하다
cause the aircraft to bank 항공기를 뱅크시키다
link together 결합하다
so that so 이전이 원인, 이유가 되어 that 이하가 되다. that 이하를 위해 so 이전이 되다.
opposite 반대측의

〈해 석〉
보조 날개는 조종간에 가해진 횡방향의 힘에 따라 작동한다. 조종간을 우측 방향으로 움직이도록 가해진 힘은 우측의 보조 날개를 들어올리고 좌측의 보조 날개를 내리게 하여 항공기를 우측으로 뱅크시킨다. 보조 날개는 한쪽 보조 날개가 내려갔을 때 반대측의 보조 날개가 올라가도록 조종 케이블로 링크 결합되어 있다.

> **7** The elevators are the angle-of-attack control. When back pressure is applied on the control stick, the tail lowers and the nose rises, thus increasing the wing's angle of attack and lift.

〈단어 연구〉
　angle-of-attack 받음각
　back pressure 후방으로의 힘
　the tail (of airplane) (비행기의) 꼬리 날개
　the nose (of airplane) (비행기의) 기수
　rise 올라가다(↔descend)

〈해 석〉
승강타는 받음각을 조종하는 것이다. 조종간에 후방으로의 힘이 가해지면 미부가 내려가고 기수가 올라가는데 이렇게 해서 날개의 받음각 및 양력을 증가시킨다.

---

8 The rudder controls movement of the airplane about its vertical axis. This is the motion called yaw. When the rudder is deflected to one side, it protrudes into the airflow, causing a horizontal force to be exerted in the opposite direction.

〈단어 연구〉
　vertical axis 수직축
　yaw 편진동
　deflect 편향시키다, 빗나가게 하다
　protrude into the airflow 기류중으로 내밀다(캠버를 만들다)
　exert (힘 등을) 쓰다

〈해 석〉
방향타는 상하축 주위로 항공기의 움직임을 조종한다. 이것은 편진동이라고 하는 운동이다. 러더가 한쪽으로 움직이면 기류중으로 나와(캠버를 만들어) 반대 방향으로 움직이는 수평력을 만들어낸다.

---

9 To offset the forces that tend to unbalance an aircraft in flight, ailerons, elevators, and rudders are provided with auxiliary controls known as tabs. These are small, hinged control surfaces attached to the trailing edge of the primary control surfaces.

14 제1장 Aircraft General

⟨단어 연구⟩
offset(something that compensate for something else) 상쇄하다
tend to   ~하는 경향이 있다
unbalance 평형을 잃게 하다
be provided with   ~가 설치되어 있다
hinged control surfaces 힌지가 달린 조종면

⟨해 석⟩
비행중에 항공기의 균형을 잃게 하는 힘을 상쇄하기 위해 보조 날개, 승강타 및 방향타에는 탭으로 알려져 있는 보조 조종 장치가 설치되어 있다. 이들 탭은 주조종면의 트레일링에이지에 부착된 작은 힌지가 달린 조종면이다.

---

10 To trim means to correct any tendency of the aircraft to move toward an undesirable flight attitude. Trim tabs control the banance of an aircraft so that it maintains straight and level flight without pressure on the control column, control wheel, or rudder pedals.

⟨단어 연구⟩
to trim 트림을 하는 것
to correct 수정하는 것(to set or make right ; remove the errors or faults from ; The new glasses corrected his eyesight) tendency to ~하는 성향
move toward ~(방향)으로 움직이다, 향하다
undesirable 바람직하지 않은
flight attitude 비행 자세
balance 균형
so that ~가 되도록
control column 조종간
control wheel 조종 휠

⟨해 석⟩
트림을 취한다는 것은 바람직하지 않은 비행 자세를 향하려고 하는 항공기의 경향을 수정하는 것을 의미한다. 트림 탭은 조종간, 조종휠 또는 러더 페달에 힘을 가하지 않고 직선 수평 비행을 유지하도록 항공기의 균형을 조종한다.

> [11] High-lift devices are used in combination with airfoils in order to reduce the takeoff or landing speed by changing the lift characteristics of an airfoil during the landing or takeoff phases.

〈단어 연구〉

high-lift device 고양력 장치
in combination with ~와 조합되어
airfoil 익형
in order to ~하기 위해
lift characteristics 양력 특성
phase 단계

〈해 석〉

고양력 장치는 착륙 또는 이륙의 단계에서 날개의 양력 특성을 바꿈으로서 이륙이나 착륙 속도를 줄이도록 날개와 조합해서 사용된다.

16 제1장 Aircraft General

1-2. Helicopter

1) General

> **1** It is known that before the days of the Roman empire the Chinese constructed "Chinese Tops". The top consisted of a propeller on a stick which was spun between the hands. The "Chinese Tops" probably represented the world's first helicopter.

〈단어 연구〉

   it is known that ~라는 것이 알려져 있다
   the days of the Roman empire 로마 제국의 시대
   Chinese Tops 중국 팽이 (도르래)
   stick 가늘고 긴 나무, 축
   spun spin(회전시키다)의 과거분사
   represent 표현하다, 의미하다

〈해 석〉

   로마 제국 시대 이전에 중국인이 「중국 팽이」를 만들었던 것으로 알려져 있다. 팽이는 양손 사이에서 회전되는 축에 프로펠러가 달린 것이었다. 「중국 팽이」는 아마 세계 최초의 헬리콥터였을 것이다.

> **2** The concept of the helicopter has been dreamed of for hundreds of years. The first recorded drawings of such a machine were made by Leonardo da Vinci in 1483.

〈단어 연구〉

   concept 개념
   dream of 꿈과 같은 것을 생각한다
   for hundreds of years 수백년 동안
   drawing 도면
   Leonardo da Vinci 레오나르도 다빈치(1452~1519 : 이탈리아 전성기 르네상스 시대의 화가, 조각가, 건축가, 과학자)

〈해 석〉

헬리콥터의 개념은 수백년간 꿈꾸어져 왔었다. 이러한 기계가 기록되어 있는 최초의 도면은 1483년 레오나르도 다빈치에 의해 만들어졌다.

---

3 Dr. Heinrich Focke, Germany, built a successful machine using two rotors, side by side, rotating in opposite directions. The rotors were inclined slightly inward to provide dihedral stability, just as is done with the wings of fixed-wing aircraft.

〈단어 연구〉

Dr. Doctor의 약자
Heinrich Focke 하인리히 포케(인명)
successful 성공한
rotor (헬리콥터의) 회전 날개, 로우터
side by side (~와) 나란히
in opposite directions 반대 방향으로
incline 경사시키다
inward 내측으로
dihedral stability 상반각 안정성
just as is 마치 ~인 것처럼

〈해 석〉

독일인 하인리히 포케 박사는 나란히 반대 방향으로 회전하는 두개의 로우터를 사용한 기계를 만들었다. 마치 고정익 항공기의 날개에서와 같은 상반각 안정성을 얻기 위해 로우터는 약간 내측으로 경사져 있었다.

---

4 At the present time it enjoys an unprecedented popularity in many areas which include corporate, agriculture, construction, petroleum support, and ambulance service.

〈단어 연구〉

at the present time 현재에는
enjoy 향유하다

18　제1장　Aircraft General

```
unprecedented  전례 없는
popularity  인기
corporate  법인
agriculture  농업
construction  건설 공사
petroleum support  석유 지원
ambulance service  구급 사업
```

〈해 석〉

현재에는 법인, 농업, 건설 공사, 석유 지원 및 구급 사업을 포함한 많은 분야에서 전례 없는 인기를 누리고 있다.

> ⑤ This helicopter may be found in almost all areas of the world performing various mission. A profile of a Hughes 500 is shown in Fig.1-5.

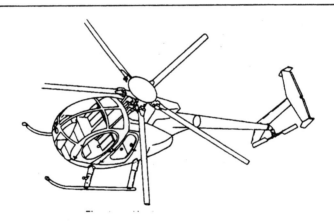

그림 1-5 Hughes 500

〈단어 연구〉

```
in almost all area  대부분의 모든 지역에서
perform  실행하다
mission  임무
profile  외관도
```

〈해 석〉
이 헬리콥터는 세계의 대부분의 모든 지역에서 다양한 임무를 수행하고 있는 것을 볼 수 있다. Hughes 500의 외관도를 그림 1-5에 나타낸다.

2) Principles of Flight

1️⃣ The helicopter as we know it today is a complex aircraft capable of flight maneuvers of hover, vertical, forward, backward, and sideward flight.

〈단어 연구〉
　today 오늘, 오늘날
　complex 복잡한
　capable of ～가 가능한
　maneuver (비행기 등의) 정교한 조종
　hover (헬리콥터가) 하버링하다

〈해 석〉
우리들이 오늘날 알고 있는 헬리콥터는 하버링, 수직, 전진, 후진 및 횡진의 비행 조종이 가능한 복잡한 항공기이다.

2️⃣ During hovering flight in a no-wind condition, the tip-path plane is horizontal, that is, parallel to the ground. Lift and thrust act straight up ; weight and drag act straight down. The sum of the lift and thrust forces must equal the sum of the weight and drag forces in order for the helicopter to hover.

〈단어 연구〉
　no-wind 무풍의
　horizontal 수평의
　that is 즉, 다시 말해서
　parallel to ～에 평행한
　straight up 곧바로 윗쪽으로

제1장 Aircraft General

   straight down 곧바로 아랫쪽으로
   sum 합계
   equal ~와 같은
   in order for the helicopter to hover 헬리콥터가 하버링하기 위해

〈해 석〉
   무풍의 조건에서 하버링하는 동안 로우터 끝이 만드는 회전면은 수평, 즉 지상에 평행하다. 양력 및 추력은 똑바로 상방으로 작용하고 중량 및 항력은 똑바로 하방으로 작용한다. 양력 및 추력의 합계는 헬리콥터가 하버링하기 위해서는 중량 및 항력의 합계와 같아야 한다.

> **3** If lift and thrust are less than weight and drag, the helicopter descends vertically : if lift and thrust are greater than weight and drag, the helicopter rises vertically.

〈단어 연구〉
   less than ~보다 적은
   descend 강하하다
   rise 상승하다

〈해 석〉
   만약 양력과 추력이 중량과 항력보다 적을 경우는 헬리콥터가 수직으로 하강하고, 만약 양력과 추력이 중량과 항력보다 클 때는 헬리콥터가 수직으로 상승한다.

> **4** The force that compensates for torque and keeps the fuselage from turning in the direction opposite to the main rotor is produced by means of a tail rotor(an auxiliary rotor) located on the end of the tail boom.

〈단어 연구〉
   compensate 보충하다(to counterbalance ; offset ; be equivalent to 비슷한 말로는 counterpoise, countervail, remunerate, reward, pay, atone)
   torque 토큐
   keep~ from~ (~하는 것을) 방해하다, 막다

by means of ~에 의해
auxiliary rotor 보조 회전 날개
tail boom 테일 붐

〈해 석〉

토큐를 보충하고 동체가 메인 로우터와 반대 방향으로 회전하는 것을 막는 힘은 테일 붐의 끝에 위치한 테일 로우터(보조 회전 날개)에 의해 만들어진다.

> 5 This auxiliary rotor, generally referred to as a tail rotor, produces thrust in the direction opposite to torque reaction developed by the main rotor. Foot pedal in the cockpit permit the pilot to increase or decrease tail-rotor thrust, as needed, to neutralize torque effect.

〈단어 연구〉

torque reaction 토큐의 반작용
developed by ~에 의해 발생한
as needed 필요에 따라
neutralize ~의 효력을 없애다, 중화시키다

〈해 석〉

이 보조 회전 날개는 일반적으로 테일 로우터라고 하며 메인 로우터에 의해 생긴 토큐의 반작용과 반대 방향으로 추력을 발생한다. 조종실의 푸트 페달은 조종사가 토큐 효과를 없애는데, 필요에 따라 테일 로우터의 추력을 증가시키거나 또는 감소시키거나 할 수 있게 한다.

3) Helicopter Components and Function

> 1 The main rotor is the wing of the helicopter. In addition to the normal stresses placed on the wing, there are stresses imposed on the rotor system by centrifugal force.

제1장 Aircraft General

〈단어 연구〉
in addition to ~에 더하여, ~에 추가로
normal stress 통상의 응력
place on ~에 걸리다, ~에 놓다
impose on ~에 부과하다
centrifugal force 원심력

〈해 석〉
메인 로우터는 헬리콥터의 주익이다. 날개에 걸리는 일반적인 응력에 더하여 원심력에 의해 로우터 시스템에 작용하는 응력이 있다.

> **2** There are three fundamental types of main rotor system : fully articulated rotors, semirigid rotors, and rigid rotors.

〈단어 연구〉
fundamental 기본적인
articulated 관절형
semirigid 반관절형
rigid 무관절형

〈해 석〉
메인 로우터 시스템에는 3개의 기본형이 있다 : 전관절형 로우터, 반관절형 로우터 및 무관절형 로우터이다.

> **3** The engine of a helicopter must operate at a relatively high speed while the main rotor turns at a much lower speed. This speed reduction is accomplished through reduction gears in the transmission system.

〈단어 연구〉
relatively 비교적
reduction gear 감속 기어
transmission 감속 장치, 트랜스미션

〈해 석〉

메인 로우터가 아주 저속도로 회전하는 반면, 헬리콥터의 엔진은 비교적 고속도로 작동한다. 이 회전 속도의 감속은 트랜스미션 시스템의 감속 기어를 통하여 이루어진다.

4️⃣ Because of the much greater weight of a helicopter rotor in relation to the power of the engine than the weight of a propeller in relation to the power of the engine in an airplane, it is necessary to have the rotor disconnected from the engine to relieve the starter load.

〈단어 연구〉

much (최상급을 수식해서) 아주 크게
in relation to ~에 관련하여
have the rotor disconnected from 로우터를 ~에서 분리해두다
relieve 덜어주다(to ease or alleviate(pain, distress etc) 비슷한 말로는 mitigate, lessen, aid, help, assist, support, sustain, 반대말로는 intensify)
starter load 시동기 부하

〈해 석〉

비행기 엔진의 출력에 대한 프로펠러의 중량에 비해서 헬리콥터의 경우는 로우터의 중량이 엔진 출력보다 훨씬 더 크기 때문에 시동지 로우터를 엔진으로부터 분리시켜서 시동기의 부하를 덜어주는 것이 필요하다.

〈단어 연구〉

5️⃣ For this reason, it is necessary to have a clutch between the engine and rotor. The clutch allows the engine to be started and gradually assume the load of driving the heavy rotor system.

---

for this reason 이 이유 때문에
clutch 클러치
gradually 서서히
assume (역할 등을) 담당하다

〈해 석〉

이 이유 때문에 엔진과 로우터 사이에 클러치를 장치하는 것이 필요하다. 클러치는 엔진이 시동되는 것을 가능하게 하고 무거운 로우터 시스템을 구동하는 부하를 서서히 담당한다.

1-3. Jet Engine

1) Jet Propulsion

1 No one knows who first discovered the jet-propulsion principle, but the honor is sometimes given to a man named Hero, who lived in an Alexandria, Egypt, about 150 B.C. He invented a toy whirligig turned by steam, as illustrated in Fig.1-6 and called his invention an Aeolipile, but apparently he did not discover any very useful purpose for his discovery. The historical records are not very definite in describing the Aeolipile. If it resembled the picture in Fig.1-6 it was a primitive form of a jet or reaction engine.

〈단어 연구〉

no one knows 아무도 모른다
jet-propulsion principle 젯트 추진 원리
honor 명예
whirligig 회전 완구
illustrate 도해하다
apparently 명백히
resemble ~을 닮다
primitive 원시적인

그림 1-6

제1장 Aircraft General

〈해 석〉

　젯트 추진 원리를 누가 발견했는지 아는 사람은 없지만 때때로 그 명예가 서역 기원전 150년경 이집트의 알렉산드리아에 살고 있던 히로라는 이름의 남자에게 주어지는 경우가 있다. 그는 그림 1-6과 같은 증기로 움직이는 회전 완구를 발명하여 그것을 아에오리파일이라고 불렀다. 그러나 그가 자기의 발명을 유용하게 쓸 용도를 발견하지 못한 것은 명백하다. 이 역사의 기록은 아에오리파일을 설명하기에는 약간 애매한 점이 있다. 만약 그것이 그림 1-6의 그림과 비슷하다고 하면 그것은 젯트 또는 반작용 엔진의 초기의 형이라고 할 수 있을 것이다.

---

2 The mechanical arrangement of the gas turbine engine is simple, for it consists of only two main rotating parts, a compressor and a turbine, and one or a number of combustion chambers.

〈단어 연구〉

　mechanical arrangement 기계적 배치
　rotating parts 회전 부품
　compressor 압축기
　turbine 터빈
　a number of 일군의
　combustion chamber 연소실

〈해 석〉

　가스 터빈 엔진의 기계적 배치는 간단하다. 즉 불과 두개의 주요한 회전 부품인 압축기와 터빈 그리고 1개 또는 일군의 연소실로 구성되어 있다.

---

3 A significant feature of the gas turbine engine is that a separate section is devoted to each function, and all function are performed simultaneously without interruption.

〈단어 연구〉

　significant feature 커다란 특징
　separate section 개개의 섹션
　is devoted to ~을 전담하다
　simultaneously 동시에

1-3. Jet Engine

without interruption 연속해서

⟨해 석⟩
가스터빈 엔진의 큰 특징은 개개의 섹션이 각각의 기능을 전담하고 모든 기능이 연속해서 동시에 행해지는 것이다.

4 Jet propulsion is a practical application of Sir Isaac Newton's third law of motion which states that, for every force acting on a body there is an opposite and equal reaction.

⟨단어 연구⟩
jet propulsion 젯트 추진력
practical 실제적인
Sir Isaac Newton 아이작 뉴톤

⟨해 석⟩
젯트 추진력은 「물체에 작용하는 모든 힘에는 반대 방향으로 같은 크기의 반작용이 있다」는 것을 주장한 아이작 뉴톤의 운동의 제3법칙의 실제적인 응용이다.

5 The compressor section of the turbine engine has many functions. Its primary function is to supply air in sufficient quantity to satisfy the requirements of the combustion burners.

⟨단어 연구⟩
in sufficient quantity 충분한 양으로
combustion burner 연소기

⟨해 석⟩
터빈 엔진의 압축기 섹션은 많은 기능을 갖고 있다. 그 주 기능은 연소기의 요구를 만족하는데 충분한 양의 공기를 공급하는데 있다.

28 제1장 Aircraft General

> **6** Two principal types of compressors currently being used in turbojet aircraft engine are centrifugal flow and axial flow.

〈단어 연구〉
  principal 주된(1.first or highest, as in rank, importance, or value, 2.a chief or head)
  currently 현재
  centrifugal flow (공기가) 원심력 방향으로 흐르는 형태
  axial flow (공기가) 축 방향으로 흐르는 형태

〈해 석〉
  터보 젯트기의 엔진에 현재 사용되고 있는 두개의 주된 압축기의 형은 원심형과 축류형이다.

> **7** The centrifugal flow compressor is a single or two stage unit employing an impeller to accelerate the air and a diffuser to produce the required pressure rise.

〈단어 연구〉
  stage 단
  employ 사용하다
  impeller 임펠러
  diffuser 디퓨져
  pressure rise 압력 상승

〈해 석〉
  원심형 압축기는 공기를 가속하기 위한 임펠러와 요구되는 압력 상승을 발생시키는 디퓨저를 사용하는 1단 또는 2단의 장치이다.

> **8** The axial flow compressor is a multi-stage unit employing alternate rows of rotating(rotor) blades and stationary(stator) vanes, to accelerate and diffuse the air until the required pressure rise is obtained.

〈단어 연구〉
　　multi-stage 많은 다단
　　alternate rows of ~와 반복되는 열
　　rotor blade 동익
　　stationary 정지한
　　stator vane 정익
　　accelerate and diffuse 가속해서 확산하다(압축하는 것에 상당한다)

〈해 석〉
축류형 압축기는 로우터 브레이드(동익)와 스테이터 베인(정익)이 반복되는 열을 이용한 장치로서 필요한 압력 상승이 얻어질 때까지 공기를 가속하여 확산한다.

---

⑨ A single-spool compressor consists of one rotor assembly and stators with as many stages as necessary to achieve the desired pressure ratio and all the airflow from the intake passes through the compressor.

〈단어 연구〉
　　single-spool compressor 단축 압축기
　　as many stages as necessary to ~하는데 필요한 만큼의 단수
　　intake (공기의) 흡입구

〈해 석〉
단축 압축기는 원하는 압력비를 얻는데 필요한 만큼의 단수를 가진 하나의 로우터 어셈블리와 정익으로 되어 있으며 흡입구로부터의 모든 공기 흐름이 압축기를 통과한다.

---

⑩ The primany function of the combustion section is to burn the fuel/air mixture thereby adding heat energy to the air. The location of the combustion is directly between the compressor and the turbine sections.

〈단어 연구〉
　　fuel/air mixture 연료/공기 혼합기
　　thereby 그것에 의해

30 제1장 Aircraft General

heat energy 열 에너지
directly 똑바로

〈해 석〉

연소부의 주된 기능은 연료/공기 혼합기를 연소시키는 것인데 그것에 의해 공기에 열에너지를 부가한다. 연소부의 위치는 압축기와 터빈부 사이에 있다.

---

⑪ The turbine transforms a portion of the kinetic(velocity) energy of the exhaust gases into mechanical energy to drive the compressor and accessories. This is the sole purpose of the turbine and this function absorbs approximately 60 to 80% of the total pressure energy from the exhaust gases(Fig.1-7).

〈단어 연구〉

transform 변화시키다
kinetic energy 운동 에너지
exhaust gases 배출 가스
mechanical energy 기계 에너지
accessories 보기류
sole purpose 유일한 목적
absorb 흡수하다

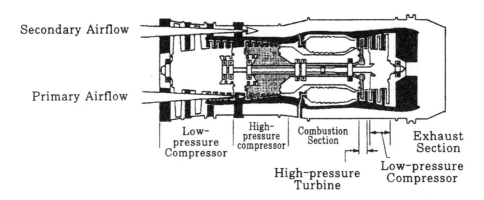

그림 1-7 Jet Engine

1-3. Jet Engine

〈해 석〉

터빈은 배출 가스의 운동(속도) 에너지의 일부를 기계적 에너지로 변화해서 압축기 및 보기를 구동한다. 이것은 터빈 고유의 목적으로 이 기능에서는 배출 가스로부터 전압력 에너지의 약 60~80%를 흡수한다(그림 1-7).

2) Engine Thrust Setting

> **1** One of the engine variables such as compressor rpm, turbine discharge pressure or engine pressure ratio, all of which vary with thrust, should be employed as an indication of the propulsive force which an engine is developing.

〈단어 연구〉

variable 변수
such as ~와 같은.
rpm 매분의 회전수(revolutions per minute의 약자)
discharge pressure 방출 압력
pressure ratio 압력비
vary with ~에 의해 바뀐다. ~에 따라 변한다
should (의무, 당연을 나타냄)~해야 한다

〈해 석〉

압축기의 회전수, 터빈 출구 압력, 또는 엔진 압력비라고 하는 추력에 따라 변화하는 엔진의 변수중 어느 것인가가 엔진이 발생하는 추진력의 지표로서 사용된다.

> **2** Although rpm is usually considered satisfactory for this purpose on centrifugal-compressor engines, turbine discharge pressure or engine pressure ratio, rather than rpm, is recommended as the operating variable for measuring the thrust output of most axial-compressor turbojet and turbofan engines.

32 제1장 Aircraft General

〈단어 연구〉
    although ~이기는 하지만
    usually 흔히
    consider 생각하다, 고안하다
    satisfactory 더할 나위 없는, 만족한
    for this purpose 이 목적을 위해
    discharge pressure 배출 압력
    rather than ~보다 오히려
    recommend 추천하다
    operating variables 작동 변수
    measure 측정하다
    most 대부분의

〈해 석〉
    원심형 압축기의 엔진에서는 회전수가 흔히 이 목적(추진력의 지표)에 충분히 맞는 것으로 생각되고 있으나, 대부분의 축류형 압축기의 터보 젯트나 터보 팬 엔진의 출력을 측정하는 계수로서는 회전수보다는 오히려 터빈 출구 압력 또는 엔진 압력비가 사용된다.

---

3 Engine pressure ratio indicators measure the pressure of the air as it enters the engine, and the gas pressure at the turbine discharge. The ratio between the two is read directly on the instrument dial. This is the engine pressure ratio.

〈단어 연구〉
    engine pressure ratio indicator 엔진 압력계, 흔히 EPR 계기라고 한다
    enter 들어가다
    ratio between the two 양자의 비
    instrument dial 계기판

〈해 석〉
    엔진 압력계는 엔진에 유입하는 공기의 압력과 터빈 출구의 가스 압력을 측정한다. 이 두개의 압력비를 계기의 눈금으로 바로 읽는다. 이것이 엔진 압력비이다.

3) Fuel Consumption

1 To enable an accurate comparison to be made between turbojet engines, fuel consumption is reduced to a common denominator, applicable to all types and sizes of turbojet engines. The term used is thrust specific fuel consumption.

〈단어 연구〉

　　enable ~하는 것을 가능하게 하다
　　comparison 비교
　　fuel consumption 연료 소비
　　is reduced to ~로 변형된다
　　common denominator 공약수
　　applicable to ~에 적용할 수 있다

〈해　석〉

　　터보 젯트 엔진을 정확히 비교하기 위해 연료 소비는 모든 타입 사이즈의 터보 젯트 엔진에 적용되는 공약수로 정리된다. 여기에 사용하는 용어가 추력 연료 소비율(TSFC)이다.

제2장 AIRCRAFT OPERATION

  항공기는 수송에 의해 그 목적이 달성되는데, 이 장에서는 항공기가 지상에서 하늘로 날아올라 일정한 비행 뒤에 다시 지상으로 되돌아오려면 항공기를 어떻게 해서 조종하고 있는가 하는 것을 중심으로 운항에 관계되는 주요한 것에 대한 영문을 소개하기로 한다.

2-1. Cockpit Operations

  항공기의 운항중에서 특히 중요한 것은 무엇보다도 조종실 내에서의 조종사의 일이다. 최근에는 젯트 항공기가 주류를 이루고 운항 업무도 복잡해졌지만 여기서는 가장 기본적인 조작으로 비행하는 소형 프로펠러 항공기를 예로 들어 그 조작에 대해 살펴보기로 하자.

1) Starting Engine

> **1** The use of an external power source for starting is recommended. With the external power source connected, it is preferable to start the airplane with the master switch "OFF". If the master switch is "ON" during the engine start, weak airplane batteries will drain off part of the current supplied by the external power source, resulting in less electrical power available for the start. After the external power source is disconnected, the master switch should be turned "ON" to supply power to electrical equipment.

〈단어 연구〉
　　external power source 외부 전원
　　preferable 나은, 바람직한(1.worthy to be preferred, 2.more desirable)
　　master switch 마스터 스위치, 주전원 스위치
　　drain off 유출시키다

제2장 Aircraft Operation

〈해 석〉

엔진을 시동하려면 외부 전원을 사용하는 것이 좋다. 외부 전원을 접속했으면 마스터 스위치를 끄고 엔진을 시동하는 것이 바람직하다. 만약 엔진 시동중에 마스터 스위치가 들어가 있으면, 비행기에 탑재되어 있는 밧데리가 약할 경우 외부 전원에서 공급되고 있는 전류의 일부를 유출시켜, 그 결과 시동에 사용할 수 있는 전력이 감소하게 된다. 외부 전원을 분리한 후 마스터 스위치를 넣어 전기 기기로 전력을 보내게 한다.

2) Taxiing

1 At some time early in the taxi run, test the brakes and note any unusual reaction, such as uneven braking. If brake operation is not satisfactory, return to the tie-down location and correct the malfunction. The operation of the turn-and-bank indicator and directional gyro also should be checked during taxiing.

〈단어 연구〉

taxiing 항공기가 자력으로 지상 주행하는 동작
unusual 이상한(not usual, common or ordinary, 비슷한 말로는 extraordinary, rare, strange, remarkable, singular, curious, gueer, odd)
uneven 고르지 못한(not level or flat ; rough ; rugged, irregular ; varying ; not uniform, not equally balanced or symmetrical)
malfunction 고장(failure to function properly)
turn-bank indicator 선회계
directional gyro 정침의, DG라고 약하기도 한다

〈해 석〉

택싱으로 주행하고 있는 동안 조기에 브레이크 테스트를 하여 브레이크의 고르지 못한 작동이 있는지 주시한다. 브레이크 작동이 만족스럽지 않을 경우는 계류 지점으로 돌아가 그 고장을 수리한다. 선회계와 정침의도 택싱중에 점검해야 한다.

3) Before Takeoff

2-1. Cockpit Operations

> ① Because just prior to takeoff you usually are distracted by other important duties such as appraising the field length, communication with the tower, setting up navigation radio frequencies, and observing other traffic, you should use the check list in the airplane. Otherwise an important check item may be overlooked.

〈어구의 풀이〉

distract 흩뜨리다, 괴롭히다(to draw away or divert, as the mind or attention, ex)The music distracted him from his work)
appraise 확인하다, 감정하다(1.to estimate the nature or value of, 2.to estimate the monetary value of ; assess, ex)He appraised the diamond ring at $500)
set up 조절하다
otherwise 그렇지 않으면
overlook 간과하다

〈해 석〉

이륙 직전에는 일반적으로 활주로 길이의 확인, 관제탑과의 교신, 항법 무선 주파수의 조절, 다른 비행기의 감시 등 중요한 업무에 주의가 산만해질 수 있으므로, 조종사는 기내에 탑재되어 있는 첵크 리스트를 사용해야 한다. 그렇지 않으면 중요한 점검 항목을 간과하는 경우가 있다.

4) After Takeoff

> ① Power reduction will vary according to the requirements of the traffic pattern, surrounding terrain, gross weight, temperature, and engine condition. However, a normal "after-takeoff" power setting is 24 inches of manifold pressure and 2,450 rpm.

〈단어 연구〉

surrounding terrain 주위의 지형
gross 전체(without deductions, opposed to net)

38 제2장 Aircraft Operation

engine manifold pressure 엔진의 흡입관 압력

〈해 석〉

출력을 줄이는 것은 비행 형태, 주위의 지형, 기체 총중량, 온도 및 엔진의 상태 등의 조건에 따라 변한다. 그러나 정상적인 「이륙 후」의 출력은 흡입관 압력이 24 in이고 회전수는 분당 2,450rpm이다.

5) Climb

1️⃣ A climbing at 24 inches of manifold pressure, 2,450 rpm (approximately 75% power) and 120 to 140 mph is recommended to save time and fuel for the overall trip. In addition, this type of climb provides better engine cooling, less engine wear, and more passenger comfort due to lower noise level.

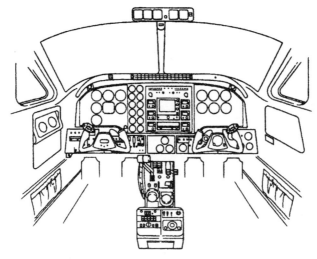

그림 2-1 Cockpit

〈단어 연구〉

save 절약하다
overall 전체(1.from one extreme limit of a thing to the other, 2. covering or including everything)

〈해 석〉

흡입관 압력이 24 in, 회전수가 분당 2,450(약 75%의 출력), 그리고 시간당 120~140마일의 속도로 상승하여 시간과 연료를 절약하는 것이 바람직하다. 또한 이 상승 요령에서는 엔진이 잘 냉각되고 엔진의 마모가 적어지며, 소음이 적기 때문에 승객의 쾌적성이 향상된다.

6) Cruise

1 Half an hour after takeoff we reach our cruising altitude of eighteen thousnad feet. I'm not doing any actual flying now ; the automatic pilot keeps us on course. But the copilot and I watch the instruments carefully.

〈해 석〉

이륙 30분 후 항공기는 순항 고도 18,000ft에 달한다. 나는 현재 실제로 조종을 하고 있지 않는다. 자동 조종 장치가 기체를 비행 경로상으로 유지하고 있기 때문이다. 그러나 부조종사와 나는 주의깊게 계기를 주시하고 있다.

2 The airplane that I fly also have two radar sets. One probes ahead 150 miles, the other sweeps a two-thousnad-foot circle around and below the plane. If anything lies in the plane's path, whether it be a mountain or another plane, light flashes and a warning bell rings.

〈단어 연구〉

probe 탐사하다, 찾다
sweep 주사하다, 쓸어버리다, 청소하다

〈해 석〉

내가 조종하고 있는 비행기는 또 두 셋트의 레이더 장치가 있다. 하나는 전방 150mile을 탐사하고 또 하나는 2,000ft의 원으로 비행기의 주위와 하방을 주사하는데, 만약 비행 경로상에 산이나 다른 비행기가 있을 때는 라이트가 점등하고 경보 벨이 울리게 되어 있다.

7) Approach

① We receive approval to go down to fourteen thousnad feet. It's a rule of the air that all westbound planes fly at even altitudes. planes going eastward are given the odd-numbered altitudes. Thus there is always a thousand feet between any two planes whose paths may cross.

〈해 석〉

우리들은 14,000ft까지 내려와도 좋다는 허가를 받는다. 서방으로 가는 비행기는 모두 짝수치의 고도를 비행하는 것이 항공 교통의 규칙이다. 동쪽으로 가는 비행기는 홀수치의 고도가 주어진다. 이렇게 항상 코스가 교차하고 있는 2대의 비행기 사이에는 1,000ft의 간격이 있게 되어 있다.

8) Before Landing

① In view of the relatively low drag of the extended landing gear and the high allowable gear down speed (160 mph), the gear should be extended before entering the traffic pattern. This practice will allow you more time to confirm that the gear is down and locked. As a further precaution, leave the landing gear extended in go-ground procedures or traffic patterns for touch-and-go-landing.

〈단어 연구〉
    in view of　~의 점에서 생각하여
    drag　항력
    as a further precaution　더 한층 주의함으로서
    traffic pattern　(공항의) 교통 장주

〈해 석〉

랜딩기어를 내렸을 때 비교적 적은 항력과 기어를 내려도 무리가 없을 만한 빠른 속도 (시간당 160 mile)를 고려하여 기어는 공항의 교통 장주에 들어가기 전에 내려야 한

다. 이렇게 함으로써 기어가 내려와 락크가 걸린 것을 확인하는데 보다 충분한 시간을 부여한다. 또 더 한층 주의하여야 할 것은 고 어라운드 절차나 터치 앤드 고 랜딩을 위한 교통 장주에서는 기어를 내려놓아야 한다.

> ② Landing gear extension can be detected by a slight bump as the gear locks down, illumination of the gear down indicator light(green), absence of a gear warning horn with the throttle retarded below 12 inches of manifold pressure and visual inspection of the main gear position.

〈단어 연구〉
bump 충돌음
illumination 점등
retard 지체시키다(to make slow : delay the progress of(an action, progress)), 속력을 늦추다

〈해 석〉
랜딩기어의 내림은 기어가 내려와 락크될 때의 약간의 충돌음, 기어 강하 지시등의 점등(녹색), 엔진 흡입관 압력을 12in 이하로 스로틀을 줄여도 기어 경고 혼이 울리지 않는 것 및 메인 기어의 위치를 육안으로 검사함으로서 알 수 있다.

9) Landing

> ① We are given permission to land. A mile from the end of the runway I lower landing gear and wing flaps. Our speed slows sharply. We're doing 130 miles an hour as we cross the edge of the field. The wheels hit the concrete. I reverse the props. They blow air in the opposite direction and act as a brake. We taxi to a stop in front of the terminal.

〈단어 연구〉
reverse 역으로(opposite or contrary in position, direction, order, or character)

제2장 Aircraft Operation

〈해 석〉

우리는 착륙 허가를 받았다. 활주로 끝에서 1mile인 곳에서 나는 랜딩기어와 윙 플랩을 내린다. 속도는 급격히 떨어진다. 비행장 끝을 지날 때 시속 130mile로 비행하고 있다. 랜딩기어 휠이 콘크리트(활주로)에 닿는다. 프로펠러의 회전 방향을 역으로 한다. 프로펠러는 반대 방향으로 공기를 불게 하여 브레이크의 역할을 한다. 터미널 앞의 주기장으로 지상 활주해간다.

2-2. Radio Communications

1) Rules for Radio Communications

> **1** Aeronautical stations and aircraft stations must, during their compulsory operating hours, keep watch on the frequencies as specified.

〈단어 연구〉

   aeronautical station 항공국, 여기서는 항공 지상 원조 시설인 전파 발신국을 말한다
   compulosry 의무화되어 있는
   frequency 주파수
   as specified 정해진

〈해 석〉

   항공국 및 항공기국은 그들의 의무 운용 시간중에는 정해진 주파수를 계속 주시하여야 한다.

> **2** The operation of an aircraft station is authorized only while the aircraft is in flight or in preparation for flight. However this shall not apply to the cases where the station is operated for reception only and so on.

〈단어 연구〉

   in preparation 준비중
   reception 수신

〈해 석〉

   항공기국은 그 항공기가 항행중 또는 항행의 준비중에만 운용하도록 (당국에서) 허가하고 있다. 다만, 수신 만을 위해 운용할 때 등은 예외이다.

제2장 Aircraft Operation

> **3** Installations of any mobile station shall be capable, once communication is established, of changing from transmission to reception and vice versa in as short a time as possible.

〈단어 연구〉
installation 설비
mobile station 이동국
vice versa 그 반대로

〈해 석〉
이동국의 설비는 일단 통신이 개설되면 송신에서 수신으로 수신에서 송신으로 가능한 한 단시간에 전환할 수 있어야 한다.

> **4** When a station receives a call without certain that such a call is intended for it, it shall not reply until the call has been repeated and understood.

〈단어 연구〉
a call without certain 확실하지 않은 호출
such a call is intended for it 그러한 호출이 의도하는 바가(확실하지 않다)

〈해 석〉
무선국은 어떤 호출이 의도하는 바가 확실하지 않는 호출을 수신했을 때는 호출이 반복되고 또 이해될 때까지 응답해서는 안된다.

> **5** In radiotelephony, an aircraft station after reply from the station on the ground shall send its call sign.

〈단어 연구〉
in radiotelephony 무선 전화에서, 무선 통신에서
aircraft station 항공기국
its call sign 자기의 호출 부호

⟨해 석⟩
　무선 통신에서 항공기국은 지상의 국으로부터 응답을 받은 뒤 자기의 호출 부호를 송신해야 한다.

> **6** When an error has been made in transmission, the word "Correction" shall be spoken, the last correct group or phrase repeated, and then the correct version transmitted.

⟨단어 연구⟩
　in transmission 전송중
　the last correct group or phrase repeated 마지막의 바른 집합 또는 어구를 반복하여
　the correct version transmitted 바른 문자를 전송하다

⟨해 석⟩
　전송중에 실수가 생겼을 때는 「CORRECTION」이란 말을 송화한 후, 마지막의 정확한 말이나 어구를 반복하여 바른 문을 전송하여야 한다.

> **7** The distress signals indicate that a ship or other vehicle is threatened by grave and imminent danger and request immediate assistance.

⟨단어 연구⟩
　distress signal 조난 신호
　other vehicle 다른 이동체(여타 비행체)
　be threatened by ~의 위협을 받다
　grave and imminent danger 중대하고 급박한 위험
　request immediate assistance 즉각적인 구조를 원하다

⟨해 석⟩
　조난 신호는 항공기 또는 다른 이동체가 중대하고 급박한 위험에 처하여 즉각적인 지원을 요청함을 뜻한다.

> **8** All radio station must, on receiving the distress signal, immediately stop the emission of radio waves liable to cause interference to the distress traffic.

〈단어 연구〉
The emission of radio waves 전파의 발사
liable to cause interference 방해할 우려가 있는
distress traffic 조난 신호, 조난 교통(비행기, 선박 등)

〈해 석〉
모든 무선국은 조난 신호를 수신했을 때 조난 통신을 방해할 우려가 있는 전파의 발사를 즉시 중지해야 한다.

2) Radiotelephony

> **1** Numbers in Aviation Radio : Numbers are used in almost every radio call. Except for whole hundreds and thousands, all numbers are spoken by pronouncing each digit separately. The number "10", for example, is pronounced "one zero", and not "ten". The number "11,000" is pronounced "one one thousnad", not "eleven thousand". numbers that have decimal points, such as 121.1, are spoken with the word "decimal" at the proper place. The number above, for example, is pronounced "one two one decimal one". To avoid misunderstandings, some digits are pronounced in a slightly different manner from that used in conversation. There could be confusion on the radio between the numbers five and nine ; these are pronounced "fife" and "niner". The numeral "0" is called "zero", the numeral "3" is usually pronouced "tree".

〈단어 연구〉

radio call 무선에 의한 호출
pronounce 발음하다
digit (아라비아) 숫자
separately 따로따로
decimal point 소수점
avoid 피하다
misunderstanding 오해
confusion 혼란, 혼동
numeral 숫자의

〈해 석〉

항공 무선에서의 숫자 : 숫자는 대부분 모든 무선의 호출에 사용되고 있다. 백이나 천은 전체를 그대로 말하지만 모든 숫자는 각 자리의 수를 하나하나 따로따로 발음하여 말하게 된다. 예를 들면 숫자의 「10」은 「텐」이 아니고 「원 제로」로 발음한다. 숫자의 「11,000」는 「일레븐 싸우전드」가 아니고 「원원 싸우전드」로 발음한다. 121.1과 같이 소수점이 있는 숫자는 그 알맞는 장소에서 「데시멀」이라는 말을 하게 된다. 예를 들면 앞의 숫자는 「원 투 원 데시멀 원」이다. 오해를 피하기 위해 어떤 숫자는 회화에서 쓰이고 있는 것과는 약간 다른 방법으로 발음된다. 무선에서는 5나 9라는 수 사이에 혼란이 발생할 우려가 있으므로 이들은 「파이프」 및 「나이너」로 발음한다. 숫자의 「0」은 「제로우」, 숫자의 「3」은 통상 「트리」로 발음한다.
주 : 다음 표에 각 숫자의 읽는 법을 나타낸다.

숫자	발 음	숫자	발 음
0	zerou	6	siks
1	wʌn	7	séven
2	tu	8	eit
3	tri	9	náinər
4	fouwər	100	hʌ́ndrəd
5	faif	1,000	táuzənd

2 Letters in Aviation Radio : The names of letters pronounced on the radio can be difficult to understand because of the similarity in sound of many pairs of letters. To avoid the possibility of confusion, words are often substituted for the letters. A standardized system of words, one for each letter of the alphabet, has been adopted by the International Civil Aviation Organization.

〈단어 연구〉

similarity 유사
substitute 대용하다
standardized 표준화된
International Civil Aviation Organization 국제민간항공기구(통상 ICAO로 생략된다), 국제 연합의 전문 기관의 하나이다. 1944년의 국제민간항공조약(시카고조약)을 바탕으로 설립된 기관으로 국제 민간 항공의 안전하고 질서있는 발달 및 국제 항공 운송 사무의 건전, 또 적정한 운영을 도모하는 것을 사명으로 하며 여러가지 활동을 하고 있다.

〈해 석〉

항공 무선에서의 문자 : 무선에서 발음되는 문자의 호칭은 그들 문자의 조합중에 유사한 음이 많이 있으므로 이해가 곤란할 경우가 있다. 혼란이 생기는 것을 피하기 위해 문자 대신에 단어가 빈번히 사용된다. 알파벳의 각 문자에 하나씩 단어의 표준 체계가 국제 민간항공기구에서 채택되고 있다.
주 : 다음 표에 문자에 대응되는 단어와 그들 발음을 나타낸다.

문자	단 어	발음 기호	문자	단 어	발음 기호
A	Alfa	'ælfə	N	November	no'vembə
B	Bravo	'bra:'vou	O	Oscar	'ɔskə
C	Charlie	'tʃa:li	P	Papa	pa'pa
D	Delta	'deltə	Q	Quebec	ke'bek
E	Echo	'ekou	R	Romeo	'roumiou
F	Foxtrot	'fɔkstrɔt	S	Sierra	si'erə
G	Golf	gɔlf	T	Tango	'tæŋgo
H	Hotel	hou'tel	U	Uniform	'ju:nifɔ:m
I	India	'indiə	V	Victor	'viktə
J	Juliett	'dʒu:ljet	W	Whiskey	'wiski
K	Kilo	'ki:lou	X	X-ray	'eks'rei
L	Lima	'li:mə	Y	Yankee	'jæŋki
M	Mike	maik	Z	Zulu	'zu:lu:

2-2. Radio Communications

> ③ When an airplane is ready to taxi out for takeoff, it contacts Ground Control before it moves. Ground Control authorizes the movement and gives any information the pilot will need. He is advised of the runway in use, the altimeter setting, the wind direction and velocity, and the time.

〈단어 연구〉

Ground Control 그라운드 컨트롤, 지상 관제소
authorize 허가하다
runway in use 사용하고 있는 활주로
direction 방향
velocity 속도

〈해 석〉

비행기가 이륙하기 위해 주기장에서 지상 주행으로 옮길 준비가 갖추어졌을 때 움직이기 전에 그라운드 컨트롤에 연락한다. 지상 관제소는 그 이동을 허가하고 조종사가 필요로 하는 모든 정보를 준다. 조종사는 사용하고 있는 활주로, 고도계의 설정, 풍향이나 풍속 및 시간에 대한 정보를 받는다.

> ④ Another function of Ground Control is to deliver ATC clearance to pilots planning IFR flights. Sometimes the clearances are quite long and take long transmissions to deliver. The pilot reads the clearance back to Ground Control to make sure there is no misunderstanding. Since a lot of time is consumed in delivering and reading back ATC clearances, it is important that other radio operations on the airport are not disturbed. This is one of the reasons Ground Control is on a separate radio frequency.

〈단어 연구〉

deliver 전달하다
ATC 항공 교통 관제(Air Traffic Control의 약어)
IFR 계기 비행 방식(Instrument Flight Rule의 약어), 항공기의 비행 경로나 방법에 대해 항시 항공 교통 관제의 지시를 받으며 비행하는 것을 말한다

transmission 전달, 전송
consume 소비하다
disturb 방해하다

그림 2-2 Control Tower

〈해 석〉

지상 관제의 또 하나의 기능은 계기 비행 방식으로 비행하려고 하는 조종사에게 항공 교통 관제의 허가를 전하는 것이다. 가끔 그 허가가 매우 길고, 또 전송하는데 장시간이 필요할 경우가 있다. 조종사는 오해가 없음을 확인하기 위해 지상 관제소에 대해 그 허가를 반복해 읽는다. 항공 교통 관제의 전송과 반복에 장시간을 사용하므로 공항에서의 다른 무선 운용을 방해하지 않는 것이 중요하다. 이것은 지상 관제가 다른 무선 주파수를 사용하고 있는 이유의 하나이다.

5 Communications Practice 1

Douglas 44Y : Grantsville Ground Control, this is Douglas 44Y, Taxi instructions, VFR departure, over.

Ground Control : Douglas 44Y, this is Grantsville Ground. Taxi south for takeoff on runway 36. Wind 320°,14 knots, altimeter 30.12. Time 1646Z, over.

Douglas 44Y, roger.

Ground Control : Cessna 141Y, you are cleared to the ramp, over.

Cessna 141Y : Cessna 141Y, roger.

2-2. Radio Communications

⟨단어 연구⟩

taxi instruction 이동하기 위한 지상 활주의 지시, 택시의 지시
VFR 시계 비행 방식(Visual Flight Rule의 약어). 기상 상태가 어떤 일정한 기준 이상일 경우에 있어 원칙적으로 항공 교통 관제의 지시를 받지 않고 조종자의 독자의 판단으로 비행하는 것을 말한다
over 오버, 자신의 말이 끝났으며 상대방의 송화를 재촉하는 말
roger 라져, 지시사항을 알아들었고 그렇게 하겠다는 말
ramp 층 외의 주기장을 말하며 주기장 내의 항공기의 주기 위치를 스폿(spot)이라고 한다

⟨해 석⟩

무선 통신의 실제 1
더글러스 44Y : 그랜트빌 그라운드 컨트롤, 여기는 더글러스 44Y, 시계 비행 방식으로 이륙하려 한다. 택시 지시를 바란다. 오버
그라운드 컨트롤 : 더글러스 44Y, 여기는 그랜트빌 그라운드. 활주로 36에서 이륙하도록 남쪽으로 지상 주행하시오. 풍향 320°, 풍속 14kt, 기압 30.12 inHg. 그리니치 표준 시간으로 16시 46분이다. 오버.
더글러스 44Y : 더글러스 44Y, 알았다.
그라운드 컨트롤 : 세스너 141Y, 램프로 들어가도 좋다. 오버.
세스너 141Y : 세스너 141Y, 알았다.

---

6 Communication Practice 2

Beech 683V : Washington Ground Control, this is Beech 683V, leaving the runway, over.
Ground Control : Beech 683V, this is Washington Ground Control, roger. Taxi straight ahead. Hold short of runway 13, over.
Beech 683V : Beech 683V, wilco.
Ground Control : Beech 683V, this is Washington Ground Control, you are cleared to cross runway 13. Right turn on the ramp to Butler, over.
Beech 683V : Beech 683V, roger.

⟨단어 연구⟩

short of ~의 앞에서, ~은 제하고
wilco 지시에 따름(I will comply의 약자)

제2장 Aircraft Operation

〈해 석〉

무선 통신의 실제 2

비치 683V : 워싱톤 그라운드 컨트롤, 여기는 비치 683V, 활주로에서 벗어나려고 한다. 오버.

그라운드 컨트롤 : 비치 683V, 여기는 워싱톤 그라운드 컨트롤, 알았다. 그대로 곧바로 나가 활주로 13 앞에서 정지해 주기 바란다. 오버.

비치 683V : 비치 683V, 알았다.

그라운드 컨트롤 : 비치 683V, 여기는 워싱톤 그라운드 컨트롤, 활주로 13을 횡단해도 좋다. 램프로 좌선회하여 버틀러로 향해 가시오. 오버.

비치 683V : 비치 683V, 알았다.

---

7 Communication Practice 3

Cessna 141Y : Kent Tower, this is Cessna 141Y, number one, ready for takeoff, over.

Tower : Cessna 141Y, this is Kent Tower. Hold your position for landing traffic, over.

Cessna 141Y : Kent Tower, this is cessna 141Y, wilco.

Douglas 44Y : Kent Tower, this is Douglas 44Y, number two, ready for takeoff, over.

Tower : Roger, Douglas 44Y. Hold your position, over.

Douglas 44Y : Douglas 44Y, wilco.

Tower : Cessna 141Y, you are cleared for takeoff. Wind 040°, 5 knots. Douglas 44Y, you are cleared into position to hold, over.

Douglas 44Y : Douglas 44Y, roger.

Tower : Douglas 44Y, this is Kent Tower. You are cleared for take off, wind 040°, 5 knots.

Cessna 141Y : Kent Tower, this is Cessna 141Y. Request right turn out, over.

Tower : Cessna 141Y, this is Kent. Right turn approved.

Cessna 141Y : Cessna 141Y, roger.

Douglas 44Y : Kent Tower, this is Douglas 44Y. Request right turn out, over.

Tower : Douglas 44Y, this is Kent. Negative. Cleared for left turn out. Acknowledge.

Douglas 44Y : Kent Tower, this is Douglas 44Y. Roger, left turn out.

〈단어 연구〉

number one 유도로상의 선두 위치에 있는 항공기를 간단히 말할 때, 따라서 number two는 두번째의 위치에 있음을 나타낸다
negative 인가할 수 없는, no에 해당(↔affirmative : yes)
acknowledge 수신 확인의 응답을 하시오

〈해 석〉

무선 통신의 실제 3
세스너 141Y : 켄트 타워, 여기는 세스너 141Y, 1번 위치, 이륙 준비 완료, 오버.
타워 : 세스너 141Y, 여기는 켄트 타워. 착륙기가 있으니 현재 위치에서 대기하시오, 오버.
세스너 141Y : 켄트 타워, 여기는 세스너 141Y, 지시에 따르겠다.
더글라스 44Y : 켄트 타워, 여기는 더글러스 44Y, 2번 위치 이륙 준비 완료, 오버.
타워 : 알았다., 더글러스 44Y, 현재 위치에서 대기하시오, 오버.
더글러스 44Y : 더글러스 44Y, 지시에 따르겠다.
타워 : 세스너 141Y, 이륙을 허가한다. 풍향 040°, 풍속 5kt. 더글러스 44Y, 이륙 활주 개시 지점에 들어가 대기하시오, 오버.
더글러스 44Y : 더글러스 44Y, 알았다.
타워 : 더글러스 44Y, 여기는 켄트 타워 이륙을 허가한다. 풍향 040°, 풍속 5kt.
세스너 141Y : 켄트 타워, 여기는 세스너 141Y. 우비행 선회를 요구한다, 오버.
타워 : 세스너 141Y, 여기는 켄트. 우비행 선회를 허가한다.
세스너 141Y : 세스너 141Y, 알았다.
더글러스 44Y : 켄트 타워, 여기는 더글러스 44Y. 우비행 선회를 요구한다, 오버.
타워 : 더글러스 44Y, 여기는 켄트. 허가하지 못한다. 좌비행 선회를 허가한다. 수신 확인의 응답을 바란다.
더글러스 44Y : 켄트 타워, 여기는 더글러스 44Y, 알았다. 좌선회한다.

8 Communication Practice 4

Mooney 141M : Kent Tower, this is Mooney 141M, entering left downwind for runway 6, over.

Tower : Roger, Mooney 141M. Traffic is a Convair on 2 mile final. Report base leg, over.

Mooney 141M : Kent Tower, this is Mooney 141M. We have the traffic in sight. Will report base.

54   제2장 Aircraft Operation

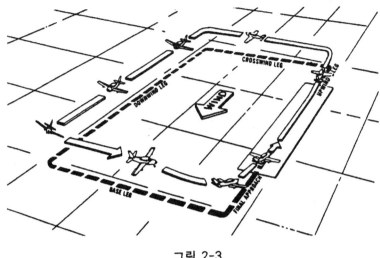

그림 2-3

⟨단어 연구⟩

   downwind 다운 윈드, 착륙 활주로와 평행이고 진행 방향과 반대인 트래픽 패턴의
   일부분을 말한다
   final 파이널, 착륙의 최종 진입을 말한다(final approach to landing의 약칭)
   base leg 베이스 레그, 활주로에 직각인 진입 최종 부분에 해당되는 트래픽 패턴의
   일부분을 말한다
   in sight 시계에 들어와, 보여

⟨해 석⟩

   무선 통신의 실제 4
   무니 141M : 켄트 타워, 여기는 무니 141M, 활주로에 착륙하기 위해 좌측 다운 윈드
   에 들어간다, 오버.
   타워 : 알았다, 무니 141M, 트래픽 패턴에는 컨베어기가 1기, 최종 2 mile 지점을 착
   륙 진입중, 베이스 레그에 들어가면 연락해 주기 바란다, 오버.
   무니 141M : 켄트 타워, 여기는 무니 141M, 콘베어기가 보이고 있다, 베이스 레그를
   통과한다.

2-3. Passenger Service

1) Information on Baggage

> ① Please do not put any valuables in your checked baggage and be sure that your baggage is locked before checking it in.

〈단어 연구〉

    valuable 귀중품, 고가의 물건
    checked baggage 위탁 수화물, 맡기는 수화물
    be sure 반드시 ~해 주십시요, 확실히 하다

〈해 석〉

    위탁 수화물에는 귀중품 등을 넣지 않도록 하시고, 맡기기 전에 반드시 자물쇠를 잠그도록 부탁드립니다.

> ② Name tags will help us to identify and expedite the return of your baggage, if it is misplaced. Does your baggage have name tags attached? If not, please use the attached tags.

〈단어 연구〉

    name tag 이름을 기입한 명찰
    identify 확인하다
    expedite 진보하다, 빠르게 하다
    misplace 잘못 두다, 분실하다
    attached tags 안내와 함께 동봉되어 있는

〈해 석〉

    이름을 기입한 명찰은 만일 수화물의 분실 사고가 발생했을 경우, 그것을 확인하거나 주인에게 빨리 돌아갈 수 있게 하는데 매우 유용합니다. 수화물에는 명찰을 부착하셨습니까? 만약 붙이지 않으셨다면 부디 이 안내와 함께 동봉되어 있는 명찰을 사용해 주십시요.

56 제2장 Aircraft Operation

> ③ Please try to minimize the amount of carry-on baggage so that you may enjoy a more comfortable flight.

〈단어 연구〉
  minimize 최소한으로 하다, 가능한 한 적게 하다
  carry-on baggage 기내 지참 수화물
  comfortable 쾌적한

〈해 석〉
  보다 쾌적한 여행을 즐길 수 있게 기내로의 휴대품은 가능한 한 최소로 하시도록 부탁 드립니다.

> ④ For the sake of safety, please store your baggage under your seat(Approximate storage dimensions are 9″×13″×23″).

〈단어 연구〉
  for the sake of ~을 위해
  store 저장하다, 넣다
  dimension 치수

〈해 석〉
  기내 지참 수화물은 안전을 위해 좌석 밑에 놓아 주십시요(놓는 장소의 크기는 대략 9 in×13 in×23 in이다).

  2) Passenger Address in Cabin

> ① Good morning, ladies and gentlemen. We welcome you aboard Chung Yeon Air Lines. This is Flight 028, bound for New York. Your flight crew is commanded by Captain Kim. This is your purser, Mr. Park. Will you please fasten your seat belt and observe the no smoking sign until it is turned off. We would like to do all we can to make your flight pleasant one, so please feel free to call on us at any time. Thank you.

〈단어 연구〉
 bound for　~로 가는, ~행
 fasten 조이다
 observe 주시하다
 no smoking sign 금연 사인
 turn off 꺼지다, 끄다
 feel free to call 자유롭게 부르다

〈해 석〉
여러분, 안녕하세요. 오늘 청연항공에 탑승해 주셔서 감사합니다. 이 비행기는 028편, 뉴욕행입니다. 오늘의 기장은 김, 저는 사무장인 박입니다. 죄송합니다만, 좌석의 벨트를 착용해 주시고 금연 사인이 없어질 때까지는 금연해 주시기 바랍니다. 여러분이 항공 여행을 즐기실 수 있도록 최선을 다하겠습니다. 용건이 있을 때는 저희들 승무원에게 언제라도 말씀해 주시기 바랍니다.

2 Ladies and gentlemen. We will be taking off in a few seconds. Please make sure that your seat belt is fastened. Thank you.

〈해 석〉
안내 말씀드립니다. 이 비행기는 곧 이륙하오니 좌석 벨트를 다시 한번 확인해 주시기 바랍니다.

3 Ladies and gentlemen. The no smoking sign has just been turned off. Your ashtray is located in the armrest of your seat. As we are still ascending, please keep your seat belt fastened. Thank you.

〈단어 연구〉
 ashtray 재떨이
 armrest 팔걸이
 ascend 상승하다
 keep belt fastened 벨트를 맨 채로 두다

58　제2장　Aircraft Operation

그림 2-4　Cabin

〈해 석〉

안내 말씀드립니다. 지금 금연 사인이 꺼졌습니다. 재떨이는 좌석 팔걸이에 있습니다. 비행기는 계속 상승중이므로 좌석 벨트는 그대로 매주시기 바랍니다.

> ❹ Ladies and gentlemen. We will be serving dinner very soon. Passengers are kindly requested to return to your seats. Thank you.

〈해 석〉

여러분에게 안내 말씀드립니다. 지금부터 식사 서비스를 하려고 하오니, 죄송하지만 서 계신 승객 여러분은 좌석으로 돌아가 주십시오.

> ❺ Ladies and gentlemen. We are sorry to announce that we have to return to the airport terminal for technical purposes. Thank you.

〈해 석〉

여러분, 대단히 죄송합니다만, 이 비행기는 기술적인 문제로 공항으로 돌아갑니다. 부디 양해해 주시기 바랍니다.

6️⃣ Ladies and gentlemen. We are expecting a little turbulence. Would you please make sure that your seat belt is fastened. Thank you.

〈해 석〉

안내 말씀드립니다. 지금부터 난기류가 예상되오니 좌석 벨트를 다시 한번 확인해 주시기 바랍니다.

7️⃣ Ladies and gentlemen. We are waiting for clearance from the Air Traffic Control Tower. We expect to land in a few minutes. Thank you.

〈해 석〉

여러분, 이 비행기는 지금 항공 관제탑으로부터 착륙 허가를 기다리고 있습니다. 조금 더 기다려 주시기 바랍니다.

8️⃣ Ladies and gentlemen. We are now approaching New York airport. Will you please fasten your seat belt. Thank you.

〈해 석〉

여러분, 곧 뉴욕 공항에 착륙합니다. 죄송하지만, 좌석 벨트를 매주시기 바랍니다.

9️⃣ Ladies and gentlemen. In New York all passengers will have to go through Public Health, Immigration and Customs. Please have your passport, Vaccination Certificate, and other entry documents ready. If you have

제2장 Aircraft Operation

> not received a landing card for entry into the United States, please contact one of your cabin attendants now. Those passengers having any fresh fruit, cut flowers or plants are requested to go through Quarantine prior to Customs. The inspection counters are located at both sides of the Customs area. Please have your passport and entry documents ready. Thank you.

〈단어 연구〉

Public Health (인간의) 검역
Immigration 입국 심사
Customs 세관
Vaccination Certificate 예방접종 증명서
entry document 입국에 필요한 서류
have~ ready ~을 준비해 주십시요
landing card 입국 카드
plants 식물류
Quarantine 검역

〈해 석〉

여러분께 안내 말씀드립니다. 뉴욕에서 검역, 입국 심사 및 세관의 수화물 검사가 있으므로 여러분의 여권, 예방접종 증명서, 그밖에 입국에 필요한 서류를 준비해 주십시요. 만약 아직 미국의 입국 카드를 받지 않으신 승객 여러분이 계시다면 지금 곧 승무원에게 말씀해 주시기 바랍니다. 또 과일, 생화 및 기타 식물류를 가지고 계신 분은 세관 검사를 받기 전에 식물 검역 검사가 있습니다. 식물 검역 카운터는 세관 검사장 내의 양측에 있습니다. 여러분의 여권과 입국에 필요한 서류를 지참해 주시기 바랍니다. 감사합니다.

2-4. Aircraft Accident and Hijacking

1) Aircraft Accidents

[1] About 16 : 53 local time, May 23, 1974, Saturn Airways Flight 14 crashed about 2.6 miles southeast of the Capital VOR, near Springfield, Illinois. Three crew members were killed. The aircraft was destroyed.
The outboard section of the left wing, including the No.1 engine, separated in flight from the remainder of the wing. The National Transportation Safety Board determines that the probable cause of the accident was the undiscovered, pre-existing fatigue cracks, which reduced the strength of the left wing to the degree that it failed as a result of positive aerodynamic loads created by moderate turbulence.

〈단어 연구〉

crash 추락하다
southeast of ~의 남동
destroy 파괴하다
separate 절단하다, 분리하다
remainder 나머지, 잔여
National Transportation Safety Board 국가운수안전위원회, NTSB로 약칭되어
　불리우는 경우가 많다
determine 단정하다
pre-existing 이전부터 있던
fatigue 피로
moderate 약간의, 알맞은
turbulence 난기류

〈해 석〉

1974년 5월 23일의 현지 시간으로 16시 30분경 새턴항공 14편이 일리노이주 스프링필드 근처 캐피탈 VOR국의 남동 2.6mile 부근에 추락했다. 3명의 승무원은 사망했고 항공기는 파손되었다. 비행중 No.1 엔진을 포함한 좌측 날개의 바깥쪽 부분이 날개의

나머지 부분에서 절단된 것이다. 미국운수안전위원회는 이 사고의 추정 원인을 이전부터 발견되지 않고 있던 균열이 좌측 날개의 강도를 약하게 한 결과 약간의 난기류에 의한 공기 역학적 하중으로 절단된 것으로 단정했다.

2 On November 3, 1973, at about 16 : 40, National Airlines, Inc., Flight 27, was crusing at 39,000 ft, 65 nmi southwest of Albuquerque, New Mexico, when the No.3 engine fan assembly disintegrated. Fragments of the fan penetrated the fuselage, the No.1 and 2 engine nacelles, and the right wing. As a result, the cabin depressurized and one cabin window, which was struck by a fragment of the fan assembly, separated from the fuselage. A passenger, who was sitting next to the window, was forced through the opening and ejected from the aircraft. The body of the passenger has not been recovered. The aircraft was landed safely at Albuquerque Int. Airport.

〈단어 연구〉

Inc. 회사(Incorporation의 약어)
nmi 노티컬 마일(nautical mile)의 약자
disintegrate 분해하다
fragment 파편
penetrate 관통하다, 뚫다
depressurize 감압하다
eject 방출하다
recover 발견하다, 복구하다
int. 국가의(International의 약어)

〈해 석〉

1973년 11월 3일 16시 40분경 내셔널 항공의 27편이 뉴멕시코주의 알바카키의 남서 65nm 지점을 39,000ft의 고도로 순항하고 있었는데, 그때 No.3 엔진의 팬 어셈블리가 떨어져 나갔다. 팬의 파편은 동체, No.1 및 No.2 엔진의 나셀, 우측 주익 내에 박히게 되었다. 그 결과 객실은 감압이 되어 팬 어셈블리의 파편이 닿은 1장의 창이 동체에서 떨어져 나갔다. 그 창 옆에 앉아 있던 승객이 그 뚫린 창으로 빨려나가 기체에서 튀어나갔다. 승객의 사체는 발견되지 않았다. 기체는 무사히 알바카키 국제 공항에 착륙했다.

2-4. Aircraft Accident and Hijacking

3 About 19:05, on January 6, 1974, Commonwealth Commuter Flight 317, an Air East, Inc., Beechcraft 99A, crashed while making an instrument approach to runway 33 at the Johnstown Cambria County Airport, Johnstown, Pennsylvania. Of the 15 passengers and 2 crew members aboard, 11 passengers and the captain were killed in the crash. The four remaining passengers and the first officer were seriously injured. While on an instrument landing system localizer approach the aircraft struck approach lights about 300 feet from the runway threshold and then crashed into an embankment about 200 feet from the threshold. Shortly before and shortly after the accident, the reported weather conditions at the Johnstown Airport consisted in part of variable 200 to 400 foot ceilings and a prevailing visibility of 2 miles in very light snow and fog.

〈단어 연구〉

　　instrument approach　계기 비행에 의한 진입
　　of the 15 passengers　승객 15인중
　　seriously injured　중상을 입은
　　threshold　경계, 문지방, 활주로단의 끝부분을 말함
　　embankment　제방, 둑
　　shortly before and shortly after　~의 직전 직후
　　consist of　~으로 구성되다
　　in part　부분적으로, 일부는(partly)
　　ceiling　구름 높이
　　prevailing visivility　주변 시정

〈해　석〉

　　1974년 1월 6일 16시 40분경 커먼웰스 코뮤터 317편 에어이스트 항공회사의 비치크래프트 99A형기가 펜실베니아주의 존스타운에 있는 존스타운 캠브리어 카운티 공항에서 활주로 33으로 계기 비행에 의한 진입 도중 추락했다. 15명의 승객과 2명의 승무원중 11명의 승객과 기장이 충돌시에 사망했다. 나머지 승객 4명과 부조종사는 중상을 입었다. 계기 착륙 장치의 로컬라이저에 의한 진입 도중, 이 항공기는 활주로의 스레시홀드에서 300ft 부근의 진입등에 충돌해서 스레시홀드에서 200ft 부근의 제방에 충돌한

것이었다. 존스타운 공항에서의 이 사고 직전 직후의 기상 상태는 200ft에서 400ft의 가변적인 운고로 조금씩 내리는 눈과 안개로 주변 시정의 2마일이었다고 보고되고 있다.

2) Hijacking

1 Arab Gunmen Hijack Egyptian Jet : Cairo(UPI)
Arab gunmen Monday hijacked an Egyptair jetliner carrying foreign and Arab tourists on a domestic flight from Cairo to Luxor. The gunmen threatened to blow up the plane unless five Arabs detained for conspiring to assassinate two dissidents Libya and South Yemeni political leaders were released. A Goverment spokesman said the Boeing 737 was carrying 95 passengers, including the hijackers, and a crew of six. He said the passengers consisted of 16 Arabs and 79 non-Arbs, mainly Japanese and French tourists. The passengers aboard the Egyptian airliner have released all women and children aboard the plane.

〈단어 연구〉
gunmen 무장 범인
threaten 협박하다
blow up 폭파하다
detain 억류하다
conspire 음모하다
assassinate 암살하다
dissident 의견을 달리하는, 반대자
political leader 정계 지도자

〈해 석〉
아랍 무장 범인이 이집트 공항의 젯트 여객기를 납치하다 : 카이로(UPI 통신)
아랍인 무장 범인이 월요일, 외국인과 아랍인의 여행객을 싣고 카이로에서 룩셀(Luxor)로 향하는 이집트 항공의 국내선 젯트 여객기를 납치했다. 무장 범인들은 2명의 의견을 달리하는 리비아 및 예멘의 정계 지도자의 암살 기도 혐의로 억류되어 있는 5명의 아랍인을 석방하지 않으면 기체를 폭파하겠다고 협박했다. 정부 대변인은 그 보잉 737기에는 납치범을 포함한 95명의 승객과 6명의 승무원이 타고 있다고 발표했다.

그 대변인은 승객중 16명은 아랍인이고 79명이 외국인인데 이들은 주로 일본인과 프랑스인 여행객이라고 발표했다. 납치된 이집트 항공의 젯트 여객기에 타고 있던 승객에는 일본인 28명이 포함되어 있었다. 7명의 아랍인으로 된 이집트 항공 젯트 여객기의 납치범들은 타고 있던 여성과 아이들을 전원 석방했다.

② Cyprus Gov't Promises Jijackers Free Passage : Larnaca, Cyprus(Kyodo-Reuter)
The Cyprus Government has promised a free passage out of the country to the three hijackers of a KLM jetliner after the men released their 83 hostages at the end of a day drama and suspense. The three men… identified as Palestinians… finally surrendered the DC-9 and their hostages at Larnaca airfield Sunday after a journey that took the twin-engined jet from the original hijack point near Nice, France, to Tunis, to Larnaca, into Israeli airspace and back to Larnaca, where it landed with no fuel left to spare. When the DC-9 eventually returned to Larnaca it only had enough fuel for one circuit of the airfield before landing. British Air Force rescue helicopters stood by in case it landed in the sea.

〈단어 연구〉

free passage 자유 통행
hostage 인질
surrender 건네주다, 항복하다
airspace 공역
stand by 대기하다

〈해 석〉

키프러스 정부 납치범에게 자유 통행 약속 : 라나카, 키프러스(교또—로이터)
키프러스 정부는 KLM 항공 젯트 여객기를 납치한 3명의 범인들에 대해 24시간의 드라마와 서스펜스 끝에 83명의 인질을 석방한 직후, 국외로의 자유 통행을 약속했다. 3명의 범인들—팔레스타인인으로 밝혀졌다—은 마침내 일요일 라나카 공항에서 DC-9 항공기와 인질들을 풀어주고 항복했다. 이들은 이전에 이 쌍발 젯트기를 최초 프랑스 니스 근처의 납치 지점으로부터 튜니스, 라나카를 거쳐 이스라엘 공역까지 갔다가 다시 라나카로 끌고 돌아갔다. 라나카에 내렸을 때는 여분의 연료가 전혀 남아 있지 않았다.

이 DC-9형기가 최종적으로 라나카로 돌아갔을 때는 착륙 전에 비행장 상공을 일주할 만큼의 연료 밖에 남아있지 않았다. 영국 공군의 구난 헬리콥터도 그 젯트기가 바다에 착수할 경우를 대비하여 대기하고 있었다.

3) Emergency Procedures

> 1 Attention, all passengers. This is your captain Kim speaking. Our aircraft took fire in engines and developed serious trouble inflight control system. Now we can neither continue our flight nor return to the airport.

〈단어 연구〉

attention 주의를 재촉할 때의 수식어
took fire in engines 엔진에 화재가 생겼다
developed serious trouble 중대한 고장이 발생했다
can neither ~ nor…… ~도 할 수 없고 ……도 할 수 없다

〈해 석〉

승객 여러분, 저는 기장인 김입니다. 이 비행기는 엔진에 화재가 발생하여 조종 장치에 중대한 고장이 발생했습니다. 따라서 이 이상 비행을 계속하는 것도 비행장으로 돌아가는 것도 불가능해졌습니다.

> 2 In the last resort, I was forced to select the best way of ditching our aircraft to secure the safety of all aboard.

〈단어 연구〉

in the last resort 최후의 수단으로써
ditching 착수하는 것
secure the safety 안전을 확보하다

〈해 석〉

최후의 수단으로 저는 승객 여러분의 안전을 확보하기 위해 최선의 방법으로서 이 비행기를 착수시킬 수단을 택하지 않을 수 없게 되었습니다.

2-4. Aircraft Accident and Hijacking

> **3** All crew members are well trained for this kind of situation, so please do not get panicky and follow instruction of crew members calmly. Your cooperation is requested.

〈단어 연구〉

trained 훈련을 받은
for this kind of situation 이런 종류의 사태에 대비하여
get panicky 혼란 상태가 된다
calmly 침착히
cooperation 협력

〈해 석〉

승무원은 전원 이런 종류의 사태에 대비하여 충분한 훈련을 받았으므로 부디 혼란 상태에 빠지지 마시고 침착히 승무원의 지시에 따라주시기 바랍니다. 여러분의 협력을 부탁드립니다.

> **4** As there may be 2 or more impacts on touchdown, all passengers should hold upper bodies as firmly as possible untill airplane comes to rest. I'll give you a warning when ditching is imminent.

〈단어 연구〉

impact 충격
touchdown 접지 또는 착수
come to rest 정지하다
warning 경고하다, 신호하다
imminent 긴급한, 절박한

〈해 석〉

착수시에는 2회 또는 그 이상의 충격이 생길 것으로 예상되므로 승객 여러분은 비행기가 정지할 때까지 상체를 가능한 한 단단히 고정하도록 해 주십시오. 착수의 시기가 가까워오면 그 때에는 신호를 하겠습니다.

68 제2장 Aircraft Operation

> **5** After airplane comes to rest in the water, you are requested to jump into the life rafts prepared. There are 6 life rafts loaded in this airplane and all people can be easily accommodated into the life rafts.

〈단어 연구〉
life raft 구명 보트
loaded 탑재되어 있는
accommodate 수용하다

〈해 석〉
비행기가 수상에 정지하면 여러분은 준비되어 있는 구명 보트에 타주십시오. 이 비행기에는 6개의 구명 보트가 탑재되어 있어 쉽게 여러분 모두를 수용할 수 있습니다.

> **6** In case emergency landing : After airplane comes to rest on the ground, you can evacuate out of the plane by means of escape slide through the exit you will be instructed to leave.

〈단어 연구〉
evacuate 피난시키다
by means of ~에 의해
escape slide 비상용 슬라이드

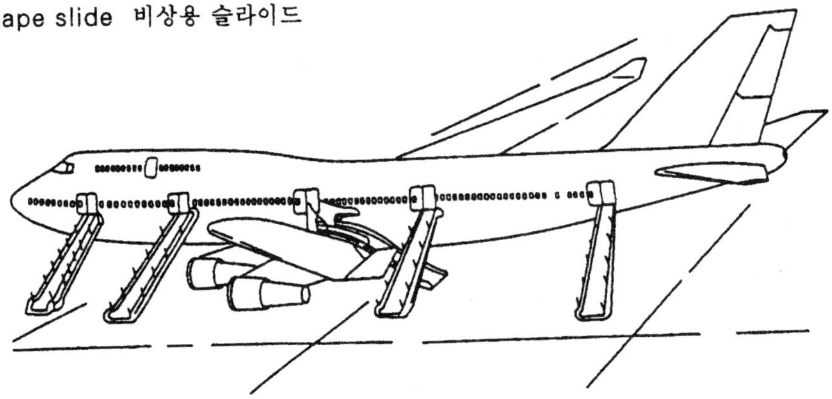

그림 2-5 Main Deck Escape System

〈해 석〉

긴급 착륙의 경우 : 비행기가 지상에 정지하면 여러분은 각기 지시되는 출구에서 비상용 슬라이드로 대피할 수 있습니다.

〈단어 연구〉

> [7] We'll ditch in about 2 minutes. One minute from touchdown. Brace yourselves for impact! Ditching! Evacuate the aircraft!

brace yourselves 자기 자신을 지키다, 분발하다. 즉 안전 자세를 취하다.

〈해 석〉

이 비행기는 앞으로 약 2분후에 착수합니다. 착수 1분전. 안전 자세를 취해 주십시오! 착수! 탈출해 주십시오!

제3장 AIRCRAFT MAINTENANCE

3-1. Maintenance Manual

1) Organization of Maintenance Manual

> 1️⃣ The material of the Maintenance Manual is organized and indexed in accordance with Air Transport Association Specification No.100, with certain departures. ATA Specification 100 envisions, each for type aircraft, a Maintenance Manual, Wiring Diagram Manual, Structural Repair Manual, Illustrated Parts Catalog, Overhual, Tool and Equipment List, Service Bulletins, and Weight and Balance Manual.

〈단어 연구〉

   material 재료, 내용
   index 색인을 붙이다, 분류하다
   Air Transport Association Specification No.100 항공기의 각 계통이나 작업 기준을 1번에서 100번까지 분류해놓은 규격으로 현재 대부분의 정비 교범은 이 분류에 따르고 있으며 ATA System Number라고도 부른다.
   with certain departures 다소의 일탈은 있어도
   Maintenance Manual 정비 기준, 정비 작업에 사용하는 기준이나 방법을 상세하게 정해놓은 것(통상 M.M.이라고 부르는 경우가 많다).
   Wiring Diagram Manual 와이어링 다이어그램 매뉴얼, 전기 계통의 배선을 도식으로 나타낸 것
   Structrual Repair Manual 스트럭츄럴 리페어 매뉴얼, 기체 구조 부분의 수리 방법을 설명해놓은 것
   Illustrated Parts Catalog 일러스트레이티드 파츠 카타로그, 부품을 부품 번호와 도면으로 상세하게 기술해놓은 카타로그
   Overhaul Manual 오버홀 매뉴얼, 항공기나 부품의 분해 손질을 하는 방법을 기술한 것
   Tool and Equipment List 툴 앤드 이큅프먼트 리스트, 공구나 장비가 기재되어 있는 리스트

제3장 Aircraft Maintenance

Service Bulletin 서비스 블레틴, 항공기의 애프터 서비스를 위해 사용하는 기술 정보의 하나로 주로 항공기 불량 상태와 그것을 수정하는 내용의 것이 많으며 항공기 제작회사에서 항공기의 사용자에게 발행된다(통상 S.B.라고 부른다).

Weight and balance Manual 웨이트 앤드 밸런스 매뉴얼, 항공기는 사용하는 회사 또는 조건이나 수리 작업 등에 의해 중량과 중심의 위치가 바뀌므로 이들을 어떤 일정한 기한에 측정하도록 의무화하고 있는데 이 측정 방법에 대해 기술해놓은 것

〈해 석〉

메인터넌스 매뉴얼의 내용은 다소의 차이는 있어도 ATA 규격 100에 따라 편성되고 분류되어 있다. ATA 규격 100은 각 기종의 메인터넌스 매뉴얼, 와이어링 다이어그램 매뉴얼, 스트럭춰럴 리페어 매뉴얼, 일러스트레이티드 파츠 카탈로그, 오버홀 매뉴얼, 툴 앤드 이큅프먼트 리스트, 서비스 블러틴 및 웨이트 앤드 밸런스 매뉴얼에 대한 기준을 제시하고 있다.

주 : ATA System Number에 따른 내용은 다음과 같다. 결번은 예비로 현재에는 사용되지 않는다. 또 System의 호칭이 기종이나 제작회사 등에 따라 약간 다를 경우가 있는데 내용에는 차이가 없다.

Sys. No. Title
1. General
6. Dimension & Area
8. Leveling & Weighing
10. Parking & Mooring
12. Servicing
21. Air Conditioning
23. Communications
25. Equipment/Furnishing
27. Flight Controls
29. Hydraulic Power
31. Instruments
33. Lights
35. Oxygen
38. Water/Waste
51. Structure
53. Fuselage
55. Stabilizers
57. Wings
71. Power Plant
74. Ignition
76. Engine Control
78. Exhaust
80. Starting

5. Time Limit/Maintenance Check
7. Lifiting & Shoring
9. Towing & Taxiing
11. Required Placard
20. Standard Practice-Airframe
22. Autopilot
24. Electric Power
26. Fire Protection
28. Fuel
30. Ice and Rain Protection
32. Landing Gear
34. Navigation
36. Pneumatic
45. CMC
49. Auxiliary Power Unit
52. Doors
54. Nacelle/Pylons
56. Windows
61. Propeler
73. Engine Fuel and Control
75. Air
77. Engine Indicating
79. Oil

2) Definitions of Maintenance Terms

> **1** Overhaul : Disassembly as recommended by the manufacturer of the component concerned or to the point where all parts subject to wear, breakage, contamination or corrosion can be adequately inspected. Replacement or rework of defective parts and replacement of seals, bearings, etcetera as may be recommended by the manufacturer and as experience dictates. Thorough cleaning, corrosion treatment, lubrication or other recommended finishing of bits and pieces. Reassembly in accordance with manufacturer's instructions. Complete test using test equipment capable of accomplishing at least minimum testing recommended by the manufacturer and of desired accuracy. Final inspection and tagging.

〈단어 연구〉

    disassembly 분해
    component 장비품, 부품
    concerned 해당하는
    subject to ～에 취약함
    wear 마모
    breakage 파손
    contamination 오염
    corrosion 부식
    adequately 적절히
    defective 결함이 있는
    dictate 지도하다, 지령하다
    thorough 완전한, 철저한
    corrosion treatment 방식 처리
    bits and pieces 소부품
    accuracy 정확도

〈해 석〉

    오버홀 : 해당 부품의 제작 회사가 추천하는 방법으로 분해하거나 마모나 파손, 오염, 부식을 받기 쉬운 부품을 모두 적절히 검사받을 수 있는 점까지 분해하는 것. 제작회사

제3장 Aircraft Maintenance

가 추천하는 방법이나 경험에 의한 방법에 따라 결함 부품을 교환 또는 수리하거나 시일이나 베어링 등을 교환하는 것. 작은 부품의 철저한 세척, 방식 처리, 급유 및 다른 적절한 방법으로 마무리를 하는 것. 제작 회사의 지시에 따라 다시 조립하는 것. 제작 회사가 추천하는 최저 한도의 시험을 할 수 있고, 또 필요한 정확도가 있는 시험 기기를 써서 완전히 시험하는 것. 최종 검사를 하고 합격품에 사용 가능의 태그를 붙이는 것.

---

2 Time Since Overhaul(TSO) : Actual logged flight time accumulated by any component subsequent to last major overhaul.

〈단어 연구〉
  accumulated 쌓아올려진, 축적된
  subsequent to 그 후의

〈해 석〉
  오버홀 이후의 시간 : 전회의 주요 오버홀 이후 해당 장비품에 있어 실제로 항공 일지에 기록된 적산 비행 시간을 말함.

---

3 Inspect : "Inspect" means an examination, visually, with or without magnifying glass, by use of magnaflux or any other accepted methods, to determine, insofar as possible, the condition, serviceability or airworthiness of an aircraft, component or unit.

〈단어 연구〉
  examination 시험, 주사
  visually 육안으로
  magnifying glass 확대경
  magnaflux 마그너플럭스, 균열 등을 조사하는 조사 방법의 하나
  accepted 일반적으로 인정되고 있는
  insofar as ～하는 한에 있어서는
  serviceability 사용 가능한 것
  unit 단위 부품, component나 assembly 등과 같은 의미로 쓰이는 경우가 많다

3-1. Maintenance Manual

〈해 석〉

검사 : 「검사」는 항공기나 장비품의 상태, 사용 가능성 또는 감항성을 결정하기 위해 확대경을 사용하거나 혹은 육안 또는 마그너 플럭스나 다른 일반적으로 인정되고 있는 방법을 써서 조사하는 것이다.

---

4 Check : The term "Check" usually means the actual operation, movement or measurement of an assembly or component to determine the operating condition of the mechanism and examination or comparison of its operational characteristics with the normal operational characteristics of the mechanism.

〈단어 연구〉

term 용어, 단어
assembly 조립 부품, unit와 마찬가지로 component와 같은 의미로 사용되는 경우가 많다.
comparison 비교
operational characteristics 작동 특성

그림 3-1 Engine Check

〈해 석〉

점검 : 「점검」이라는 용어는 메카니즘의 작동 상태가 적정한지를 결정하기 위해 장비품을 실제로 작동시키거나 움직이거나 혹은 측정하거나 하는 것, 또 그 메카니즘의 정상적

제3장 Aircraft Maintenance

인 작동 특성과 장비품의 작동 특성을 조사하거나 비교하거나 하는 것을 의미한다.

> **5** Repair : The term "Repair" is applied to the restoration of an item, aircraft or component to fully serviceable condition according to Company and KCAB regulations.

〈단어 연구〉
  Repair 수리하다
  KCAB 교통부 항공국의 약자(Korea Civil Aviation Bureau)

〈해 석〉
  수리 :「수리」라고 하는 용어는 어떤 하나 즉 항공기 또는 장비품을 회사나 교통부 항공국의 규칙에 따라 충분히 사용 가능한 상태로 수리하는 의미로 쓰인다.

> **6** Correct Discrepancies : To perform the extent and type of work required to correct any unsatisfactory condition noted during a previous inspection in order that the condition will no longer be evident at time of a subsequent inspection.

〈단어 연구〉
  correct 고치다, 수리하다
  discrepancy 고장, 불량 상태
  perform 행하다
  extent 범위
  unsatisfactory 불충분한
  previous 전회의
  in order that ~하기 위해
  no longer 더이상 ~가 아니다
  evident 명백한

〈해 석〉
  불량 상태를 고친다 : 전회의 검사에서 기록된 불량 상태를 다음 검사시에 그 상태가 더 이상 존재하지 않게 하기 위해 필요한 범위와 종류의 수리 작업을 하는 것

3-1. Maintenance Manual

> **7** Service : To perform certain predetermined maintenance work generally known to be required by the Company or recommended by the manufacturer for aircraft assemblies or systems. This term will include an inspection of pertinent characteristics during the course of the maintenance work.

〈단어 연구〉
predetermined 미리 정해진
pertinent 적절한
course 진행

〈해 석〉
서비스 : 항공기 전체 또는 각 계통에 대해 회사가 일반적으로 필요하다고 생각하거나 제조회사가 추천하는 미리 정해진 일정한 정비를 하는 것이다. 서비스에는 정비 작업진행중에 특성이 적정한지의 여부를 검사하는 것도 포함한다.

> **8** Functional Check : A check or test of the designed function and operation of a unit in the aircraft using equipment, procedure and limits established in the Maintenance Manual or other applicable manuals.

〈단어 연구〉
functional 기능의
procedure 순서, 절차, 요령
applicable 적용할 수 있는

〈해 석〉
기능 시험 : 메인터넌스 매뉴얼이나 다른 적용 가능한 매뉴얼에 정해져 있는 장비나 절차 및 제한치를 이용하여 항공기에 탑재되어 있는 장비품이 설계대로의 기능을 갖고 있는지 또 작동하는지의 여부를 점검하거나 시험하는 것을 말한다.

> **9** Bench Check : The unit shall be removed from the aircraft and checked or tested using appropriate shop

78   제3장 Aircraft Maintenance

> equipment and procedures to determine that the unit is operating within the manufacturer's tolerance with respect to performance, wear or deterioration.

〈단어 연구〉
   appropriate 적절한, 타당한
   tolerance 공차, 허용 범위
   with respect to ~의 점에 대해
   performance 성능
   deterioration 열화, 퇴화

〈해 석〉
   벤치 체크 : 항공기에서 떼내어진 장비품의 성능이나 마모 및 열화 등에 관해서 제작사의 허용 범위 이내에서 그 장비품이 작동하는지를 결정하기 위해 적절한 샵의 장비나 절차를 사용하여 점검하거나 시험하거나 하는 것이다.

3) Purpose of Aircraft Service System

> 1️⃣ The aircraft service and inspection system is designed to provide preventive periodic maintenance and inspection, and progressive overhaul.

〈단어 연구〉
   aircraft service and inspection system 항공기의 정비 및 검사 방식(정비 방식 aircraft maintenance system라고도 한다)
   progressive overhaul 프로그레시브 오버홀. 항공기 전체를 한번에 완전한 오버홀을 하려고 하면 장기간이 필요하여 경제적이지 않으므로 항공기를 몇개의 구획으로 나누어 1회에 1구획을 오버홀하는 식으로 오버홀의 시간 한계 내에서 전구획의 오버홀이 일주되도록 고안된 방식을 말한다.

〈해 석〉
   항공기 정비 및 검사 방식은 예방 정기 정비와 검사 및 계속적인 오버홀을 하도록 만들어져 있다.

3-1. Maintenance Manual 79

> ② All aircraft must be frequently inspected to obtain the highest degree of safety and efficiency during flight.

〈해 석〉
모든 항공기는 비행중 최고의 안전성과 효율을 얻을 수 있게 빈번히 검사해야 한다.

> ③ Some parts of the aircraft require more frequent inspection and service than others. Therefore, it is necessary to have different types of inspections in order that parts are inspected only as often as necessary.

〈해 석〉
항공기의 부품에는 다른 부품보다 빈번한 검사와 정비가 필요한 것이 있다. 따라서 그들 부품을 필요한 만큼 반복하여 검사하기 위해 다른 형식의 검사를 할 필요가 있다.

4) Minimum Equipment Requirement

> ① General : The following contain listings of certain instruments and accessories which may be inoperative and still not reduce the safety of operation of the aircraft below minimum limits. These lists should eliminate last minute delays if items may be "held over" to a station where both the time and equipment will be more readily available.

〈단어 연구〉
　Minimum Equipment Requirement　Minimum Carry-Over(held-over) Standard「최저 이월 기준」라고도 한다.
　inoperative 작동하지 않는
　reduce 줄이다, 저하하다
　eliminate 배제하다
　delay 지체, 늦어짐
　readily 용이하게, 간단히

제3장 Aircraft Maintenance

〈해 석〉
일반 : 다음 항목은 작동하지 않게 되어도 비행기 운항의 안전을 최저 기준 이하로는 저하시키지 않는 계기나 장비품을 열거한 것이다. 이들 목록에 기재되어 있는 항목들은 수리하는 시간과 장비의 지원을 쉽게 얻을 수 있는 기지까지 「이월」해서 최종적인 출발의 지체를 배제하는 것이다.

---

2 If after investigation of the Flight Log report a decision is made to continue the flight with unit inoperative, it must be determined that the unit is rendered free of the possibility that further extension of the damage or defect could cause a fire hazard or a mechanical failure, either internally or to its drive, mounting connections, allied assembly or system, which could endanger the safety of the aircraft in flight.

〈단어 연구〉
   investigation 조사
   render 이루다, 하다
   free of ~을 없애다
   possibility 가능성
   further extension 더 확대하는 것
   fire hazard 화재의 위험
   allied 결합되어 있는
   endanger 위험에 빠뜨리다

〈해 석〉
비행 기록부의 기록을 조사 검토한 후 작동하지 않는 유니트를 장비한 채 비행을 계속한다고 하는 결정이 된 경우는 그 유니트가 손상이나 결함이 더 커져 화재의 위험이나 내부적으로 또는 그 구동 장치, 주변에 같이 결합된 부분 및 관련된 기기나 장치에 대해 기계적인 고장의 원인이 되어 비행중의 항공기의 안전을 위험에 빠지게 할 가능성이 없게 해야 한다.

3-2. Preflight Check and Service Work Items

Preflight Check는 항공기가 스케줄에 따라 운항하고 있는 동안 가능한 지상 체류 시간(야간을 포함) 내에 실시하는 것이 보통이다. Preflight Check 에서 행해지는 작업 항목을 살펴보자.

---

[1] Accomplish the following items.
(a) Check log book.
  ① Form M-28's cleared.
  ② List items not cleared on reverse side.
(b) Perform cockpit visual check.
(c) Perform excess heat warning test.
(d) Perform empennage de-ice, controller and temperature selector operational check.
(e) Check emergency air pressure.
(f) Check operation of emergency lights.
(g) Check oxygen cylinders for pressure and accessories for condition.
  ① Pressure of cylinders.
  ② Cylinder valves and accessories for condition and security.
(h) Check hand fire extinguisher for proper safety, weigh CO_2 extinguisher if safety is broken.
(i) Check first aid kit for seal.
(j) Check security of overwing refueling filler cap.

---

1) Cockpit and Cabin

〈단어 연구〉

　Form M-28　로그 북 속의 discrepancies가 기입되는 양식명(항공사별로 다를 수 있다)

82 제3장 Aircraft Maintenance

clear 마무리하다, 처리하다
reverse side 이면
excess heat warning 엔진이나 화물실이 화재 등으로 어떤 온도 이상이 되면 경보하는 장치. over heat warning이라고도 한다.
controller 컨트롤러, 제어 장치
selector 셀렉터, 선택하는 스위치 또는 노브를 말함.
emergency air 이머젼시 에어, 유압이 저하되어 긴급히 항공기를 정지시킬 필요가 있을 때 유압 대신에 브레이크를 작동시키는 압축 공기를 말함. 일반적으로 3,000 psi이다.
emergency light 이머젼시 라이트, 비상등
oxygen 산소
cylinder 실린더, 용기를 말함
security 소화기가 단단히 장착되고 이완된 곳이 없는 상태
weigh 중량을 측정하다

그림 3-2 Cockpit

CO_2 이산화탄소
first aid kit 퍼스트 에이드 킷트, 구급 상자
overwing refueling filler cap 날개 상면의 연료 주입구의 캡

〈해 석〉

다음 항목을 완료하시오.

(a) 로그 북을 점검할 것
  ① 양식 M-28에 기록되어 있는 불량사항이 처리되어 있는지의 여부
  ② 이면에 처리되지 않은 사항이 기재되어 있는가
(b) 조종실 내 전반을 육안으로 점검할 것
(c) 오버히트 워닝을 테스트 할 것
(d) 미익 제빙 장치의 컨트롤러 및 온도 셀렉터의 작동 점검을 할 것
(e) 비상 공기의 압력을 점검할 것
(f) 비상등의 작동 점검을 할 것
(g) 산소 보틀의 압력과 부속품의 상태를 점검할 것
  ① 산소 보틀의 압력이 규정대로 있는가
  ② 산소 보틀의 밸브와 부속품의 상태가 양호한가, 또 단단히 부착되어 있는가
(h) 휴대형 소화기의 안전선을 점검하여 만약 안전선이 끊어져 있으면 탄소 가스 소화기의 중량을 측정할 것
(i) 구급 상자의 봉인이 깨져 있지 않은지 점검할 것
(j) 날개 상면의 연료 주입구가 확실히 부착되어 있는지를 점검할 것(객실의 창에서 날개 상면을 보아 확인한다)

1 Accomplish the following items.
(a) Visually inspect the aircraft from the ground.
(b) Check outside lights for operation.
(c) Check landing gear, tires, wheels and brakes for general condition and foreign objects ; clean struts.
(d) Check hydraulic power compartment for general condition and security of units, accumulators for pressure and reservoir fluid level.
(e) Reinstall all removed items.

2) Aircraft Exterior and Interior General

〈단어 연구〉
  foreign object 이물질
  accumulator 어큐뮬레이터, 축압기, 공기압에 의해 압의 맥동을 흡수하거나 유압을 축적하거나 하는 기기
  reservoir 리저버

〈해 석〉
  다음 항목을 완료하시오.

84 제3장 Aircraft Maintenance

(a) 지상에서 기체 전체를 육안으로 검사할 것
(b) 기체 외부에 있는 라이트를 점등시켜 볼 것
(c) 랜딩 기어, 타이어, 휠 및 브레이크 전반의 상태와 이물질이 없는지를 점검할 것. 또 기어의 스트러트를 깨끗이할 것
(d) 유압 동력 실내 전반의 상태, 부품류가 확실히 부착되어 있는지, 어큐뮬레이터의 공기압 및 리저버의 작동유 레벨을 점검할 것
(e) 떼어낸 것을 복구할 것

---

1. Accomplish the following items.
(a) Visually inspect pod and pylon exterior.
(b) Visually inspect the compressor inlet area ; check thrust reverser and sound supressor exterior for condition and security.
(c) Check third stage turbine buckets for condition and security.
(d) Check fire extinguisher agent container pressure. Check safety plug.
(e) Check hydraulic filters(high and low pressure) for position of indicator button.
(f) Check engine and CSD oil levels. Service as necessary. Record quantities added(in pints) in appropriate blocks below.

	No.1 Eng.	No.2 Eng.	No.3 Eng.	No.4 Eng.
Engine Oil Added				
CSD Oil Added				
Signature				

---

3) Engine, Pod and Pylon

〈해 석〉

다음 항목을 완료하시오.
(a) 엔진의 포드와 파일론 외부를 육안 검사할 것
(b) 압축기 흡입 부근 전반을 육안 검사할 것. 스러스트 리버서 및 사운드 서프렛서 외

3-2. Preflight Check and Service Work Items

면의 상태와 확실히 부착되어 있는지를 점검한다.
(c) 제3단 터빈(최후부의 터빈)의 브레이드의 상태와 확실히 부착되어 있는지를 점검한다.

	No.1 엔진	No.2 엔진	No.3 엔진	No.4 엔진
보급된 엔진 오일				
보급된 CSD 오일				
서 명				

(d) 소화제 용기압을 점검할 것. 안전 플러그를 점검할 것
(e) 유압 필터(고압과 저압용)의 인디케이터 버튼의 위치를 점검할 것
(f) 엔진과 CSD의 오일 레벨을 점검할 것. 필요에 따라 보충을 할 것. 다음의 해당란에 보급한 양(파인트로)을 기록할 것

3-3. Service Work Items

서비스 정비는 그 때문에 항공기의 운항을 계획적으로 중지하고 격납고 내에서 하는 것이다. 이 정비에서는 항공기 각각의 시스템을 담당하는 전문 정비원이 점검 및 수리를 하는데 이 전문 직종을 스킬 "skill"(최근에는 「기능」과 혼동하기 쉬우므로 직종을 트레이드(trade), 기능을 스킬이라고하는 경우가 있다)이라고 한다. 다음에 이들 스킬마다 실시하는 작업 내용을 살펴보기로 하자.

1) Inspection Skill

> 1 Installation of red tag on aircraft : Install an Aircraft Out-of-Service Tag on the captain's control wheel or alternate cockpit location.

〈단어 연구〉
  tag 태그, 표
  Aircraft Out-of-Service Tag 비행기 운항 금지표. 이 비행기는 현재 정비중이므로 운항에 사용해서는 안됨을 주지시키는 표시
  control wheel 조종간, 조종륜.
  alternate 대신의, 대체의
  location 위치, 장소

〈해 석〉
  기체에 빨간 태그를 부착함 : 기장측의 조종간 또는 조종실 내의 다른 장소에 비행기 운항 금지 태그를 부착한다.

> 2 Transfering log book items : Transfer log book items from Form M-28 pilot report to non-routine cards. Refer to Standard Practice for procedure. Review all Form M-28's items with maintenance foreman as per Standard Practice.

3-3. Service Work Items   87

〈단어 연구〉

transfer 전달하다
non-routine card 난 루틴 카드, 비정례 작업 카드를 말하며 비행중 조종사가 발견하거나 지상 정비중 발견된 고장이나 불량 사항을 기록하는 카드이다[작업자는 이 카드를 보고 수리 작업을 한다. 이것에 대해 정기적인 정비에서 미리 행하도록 정해져 있는 항목을 기재한 카드를 루틴 카드(routine card)라고 하는데 이렇게 작업 항목이 기재되어 있고 그것으로 실제로 작업을 하는 것을 작업 카드(job card 또는 work card ; 내용, 양식에 따라 job sheet 또는 work sheet라고도 함)라고 한다].
refer to 참조하다
Standard Practice 처리 기준.
procedure 순서, 절차
review 검토하다.
maintenance foreman 정비 주임
as per ~에 의해, ~에 따라

〈해 석〉

로그북의 기재 사항을 전달 : 로그북의 양식에서 조종사의 경고 사항을 난 루틴 카드에 옮겨 기록한다. 조치의 순서에 대해서는 처리 기준을 참조한다. 양식에 기재되어 있는 사항은 전부 처리 기준에 따라 정비 주임과 함께 검토할 것.

---

3 General inspection of cockpit interior : Visually inspect sliding window frames for wear and grooving ; glass for cracks, scratches, and delamination ; make operational check of sliding windows ; visually inspect interior of cockpit for security and general condition. Inspect seat belts, shoulder harness, inertia reels, seat tracks, and operating mechanism of crew seats for security and condition.

〈단어 연구〉

sliding window 슬라이딩 윈드우, 조종사의 측면에 달려 있으며, 옆으로 밀어서 여는 타입으로 비상시에는 조종사의 탈출구도 된다.
frame 틀
wear 마모
grooving 홈과 같이 되어 있는 것, 창을 닫았을 때 기체 구조측과 밀착하는 부분이 파여 있는 상태를 말한다.

88　제3장　Aircraft Maintenance

　　scratch 긁힌 자국
　　delamination 딜레미네이션, 이중 또는 삼중으로 둘러쌓여 있는 창의 패널에 일부가 벗겨져 기포가 들어간 것같은 상태가 된 것
　　shoulder harness 쇼울더 하네스, 어깨에 매는 벨트
　　inertia reel 이너셜 릴(쇼울더 하네스를 감고 있는 릴로 조종사의 좌석 등받이 내에 부착되어 있다. 조종사가 계기를 조작할 때와 같이 천천히 몸을 앞으로 숙이면 하네스가 늘어나 몸의 움직임을 따르며, 바로 앉으면 그에 따라 하네스는 감겨진다. 그러나 긴급시와 같이 급격한 전방 방향의 움직임에 대해서는 하네스는 늘어나지 않아 조종사를 좌석에 고정시키도록 작동하는 장치)
　　seat track 바닥에 고정되어 좌석이 이동하는 궤도

〈해　석〉
　　조종실 내부의 일반 점검 : 슬라이딩 윈도우의 창틀이 마모 되거나 흠상으로 되어 있지 않은지 글래스에 균열, 긁힌 자국, 디레미네이션이 없는지 육안 검사를 한다. 슬라이딩 윈도우를 작동 점검한다. 조종실 내부의 모든 부품들이 확실히 부착되어 있는지 또 전반적 상태는 어떤지 육안 검사한다. 시트 벨트, 숄더 하네스, 이너셔 릴, 시트 트랙 및 승무원용 좌석의 작동 기구가 확실히 되어 있는지 또 상태는 어떤지 검사한다.

---

❹ General inspection of cabin entrance area : Visually inspect interior of the area for security and general condition. Perform operational check of the doors. Visually inspect external surface of entrance and service door, noting specifically, tracks, pins, hooks, hinges, seals, frame, door jamb and tracks. Inspect seat belts and harness of cabin attendant's seat for security and general condition.

〈단어 연구〉
　　Cabin entrance area 객실 입구 구역
　　noting specifically 특히 주의하여
　　door jamb 도어 잼, 도어를 닫았을 때 실(Seal)과 밀착하는 부분을 말함

〈해　석〉
　　객실 입구 구역의 일반 검사 : 그 구역 내부에서 부착 상태가 확실히 되어 있는지 또 전반적인 상태는 어떤지 육안 검사한다. 도어의 작동 점검을 한다. 승객 입구 도어와 서비

스 물품 탑재 도어의 외면, 특히 도어를 닫았을 때의 도어의 상태, 도어를 열었을 때 바람 등으로 도어가 닫히지 않게 기체측에 걸어두는 후크, 힌지, 도어 시일, 프레임, 도어 잼 및 선로를 철저히 육안 검사한다. 객실 승무원용 좌석의 시트 벨트와 하네스가 확실히 되어 있는지 또 일반 상태는 어떤지 검사한다.

---

5 General inspection interior of left wheel well and main landing gear : Perform general inspection of left wheel well area, main landing gear and components, MLG fittings on rear spar and MLG beam and side brace attachment fitting at fuselage. Perform general inspection as follows ; Anti-skid control valve and box, restrictor check, priority, sequence and door sequence valves.

〈단어 연구〉

wheel well 휠 웰, 즉 바퀴 전체가 들어가는 공간을 말함(well을 bay라고 부르는 경우도 있다)
component 장비품, 기능 부품
MLG 주착륙 장치(Main Landing Gear의 약어)
fitting 피팅
rear spar 주익의 후방 스파
MLG beam 메인 랜딩기어를 부착하는 스파
side brace 사이드 브레이스, 횡방향에서 지탱하는 지주
attachment 어태치멘트, 부착
as follows 다음 사항에 대해
anti-skid 앤티스키드, 휠(Wheel)이 브레이크가 너무 들어서 회전하지 않게 되는 것(Skidding)을 막아 효율적으로 브레이크를 작동시키는 장치
restrictor 이하는 전부 맨끝의 밸브에 연결된다
restrictor check valve 유량을 제한함과 동시에 역류를 차단하는 밸브
priority valve 우선 밸브, 중요한 작동 계기에 대해 우선적으로 작동유를 흘리게 작용하는 밸브
sequence valve 순위 밸브, 행정적으로 순서대로 작동유를 흘려주는 밸브
door 기어 웰의 도어

〈해 석〉

좌측 휠 웰 내와 메인 랜딩 기어의 일반 검사 : 좌측 휠 웰 구역, 메인 랜딩 기어와 장비품, 후방 스파에 부착되어 있는 기어 피팅, 기어 빔 및 동체에 달려 있는 사이드 브레이스 부착 피팅에 대해 전반적인 검사를 한다. 다음 부품, 즉 앤티스키드 컨트롤 밸브와

제3장 Aircraft Maintenance

컨트롤 박스, 리스트릭터 첵크 밸브, 프라이오리티 밸브, 시퀀스 및 도어 시퀀스 밸브에 대해 전반적인 검사를 한다.

⑥ General inspection of wing upper and lower exterior surface : Visual check for condition and fuel leaks. Check for cracks in lower skin specially at first row of rivets inboard of Station 8 external splice plate and outboard of Station 1 splice plate. Check for leaks or loose screws in doors. Check skin access door doubler for loose rivets.

〈단어 연구〉
 row 열
 splice 겹쳐잇기
 station 스테이션, 어떤 기준선으로부터의 위치를 나타냄
 doubler 더블러, 겹쳐서 대은 판을 말함

〈해 석〉
주익 상하면의 일반 검사 : 주익의 상태와 연료 유출이 없는지 육안 점검한다. 주익 하면에서는 스테이션 8의 외부 겹처 이음판의 내측 및 스테이션 1의 중첩판 외측의 제1열째의 리벳에 대한 균열의 유무를 특히 점검한다. 액세스 도어가 부착되어 있는 스크류가 느슨하진 않은지 또 새지 않는지 점검한다. 액세스 도어의 더블러를 고정하고 있는 리벳에는 느슨한 게 없는지 점검한다.

⑦ General inspection of wing flap area : With flaps extended, spoilers up and access panels No.83, and 123 open in wing trailing edge lower surface, check the area for general condition and security noting specifically ; Flap leading edge for abrasion at seal contact, tracks, supports, carriages, and actuating mechanism ; spoilers, hinges, supports and actuating mechanism ; main landing gear beam aft face. Listed components for condition, security, fluid leaks and electrical connections. Check torque tube splines, pillow blocks and bearings for condition and security.

〈단어 연구〉

abrasion 벗겨짐, 박리
carriage 플랩이 움직일 때, 플랩과 트랙을 연결하여 트랙을 따라 상하로 움직이는 도르래 같은 장치를 말한다.
torque tube splines 동력 전달간의 결합 부분에 있는 울퉁불퉁한 형의 홈으로 이 홈이 암수로 맞물려 한쪽의 회전력을 다른쪽으로 전달하는 것
pillow block 전달간이 흔들리지 않게 하는 지지 부품을 말한다.

〈해 석〉

윙 플랩 웰 내부의 일반 검사 : 플랩은 펼치고 스포일러를 열고 주익 트레일링에이지 하면에 있는 No.83과 123의 엑세스 패널을 연 상태에서 웰 내부 전체의 일반 상태 및 부착이 확실한 지를 특히 주의해서 점검한다. 플랩 리딩에이지 시일의 접촉부분이 벗겨지지는 않았는지 트랙, 서포트, 캐리지 및 작동 기구 또 스포일러, 힌지, 서포트 및 작동 기구 또 메인 랜딩 기어 빔의 후면을 점검한다. 이들 장비품에 대해 일반적인 상태, 부착 상태, 연료나 기름 등의 유출이 없는지 또 전기 배선의 결합이 확실한 지를 점검한다. 또 토큐 튜브의 스파라인, 필로우 블럭 및 베어링의 상태나 부착이 확실한 지의 여부를 점검한다.

---

8 Operational check of horizontal stabilizer :
(a) Actuate pilot's trim switch on aileron control wheel to NOSE UP.
  ① Small wheel on pedestal will rotate.
  ② Stabilizer leading edge should move down to approx. 12°.
  ③ Pilot's flight instrument panel indicator and pedestal indicator should agree.
(b) Return stabilizer to 4° position, using pilot's trim switch.
(c) Actuate copilot's switch to the 2° position. Return stabilizer to the 4° position.
(d) While holding copilot's trim switch to NOSE UP, override by positioning captain's beep trim to NOSE DOWN.
(e) Return stabilizer 2° to 6°
(f) Indicator on pedestal and pilot's instrument panel should agree.

92 제3장 Aircraft Maintenance

〈단어 연구〉

approx 약, 대략(approximately의 약자)
pedestal 페디스탈, 조종사와 부조종사 사이에 있는 컨트롤 박스를 말함
override 오버라이드, 다른 작동을 멈추고 우선해서 작동하는 것
beep trim 비프 트림, 컨트롤 휠을 쥔 채 엄지 손가락 끝으로 움직여 미세 조종을 하는 것으로 트림 스위치의 별칭이다.

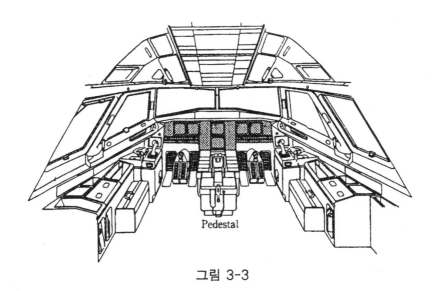

그림 3-3

〈해 석〉

수평 안정판의 작동 점검 :
(a) 조종사측의 에일러론 컨트롤 휠상에 부착되어 있는 트림 스위치를 「기수 상승」으로 작동시킨다.
   ① 페데스탈에 있는 작은 휠이 회전한다.
   ② 스터빌라이저의 리딩에이지가 약 12° 정도 하방으로 움직여야 한다.
   ③ 조종사측의 비행 계기판상의 지시와 페데스탈상의 지시가 일치해야 한다.
(b) 스터빌라이저를 조종사측의 트림 스위치를 사용하여 4°의 위치로 돌린다.
(c) 부조종사의 스위치도 2°의 위치까지 움직인다. 이어서 스터빌라이저를 4°의 위치로 되돌린다.
(d) 부조종사의 트림 스위치를 「기수 상승」의 위치로 유지한채 조종사측 비프 트림을 「기수 하강」쪽으로 움직여 오버라이드 시킨다.
(e) 스터빌라이저를 2°에서 6°까지 되돌린다.
(f) 페데스탈과 조종사측 계기상의 지시가 일치해야 한다.

9️⃣ Operational check of flight controls : Perform this operation with three inspectors working together as a team. The station for two inspectors is in the cockpit, in order to check the feel of controls while operating controls and indicators for operation. The other inspector gives directions of operation by use of telephone or walkie-talkie and will so station himself to properly observe and check the operation and direction of travel ONLY of the particular surface of surfaces being checked.

〈단어 연구〉

as a team 일조가 되어
station 위치(하다)
feel 감각, 느낌
direction 방향
properly 적절히, 바르게
observe 감시하다, 관찰하다
particular 특정한

〈해 석〉

플라잇 컨트롤의 작동 점검 : 이 작업은 3명의 검사원이 한 조가 되어 함께 하지 않으면 안된다. 2명의 검사원은 조종실에 자리를 잡고 컨트롤을 움직이고 있는 동안의 느낌과 움직임의 지시를 검사한다. 다른 검사원은 전화나 워키토키로 움직임의 방향을 가르쳐 준다. 따라서 그는 검사하고 있는 조종면중 특정한 조종면의 작동과 방향을 적절하게 관찰하고 검사할 수 있는 장소에 위치한다.

🔟 Final procedure of inspection ; Check to see that all work cards are complete and work accepted by an inspector. Check that all items not completed are of a non-airworthy nature and that these items have been properly transcribed to Form M-28 or have been deferred. Check that no test flight is required and that the aircraft is ready for service. All final inspection completed, aircraft considered airworthy and red tag removed.

94 제3장 Aircraft Maintenance

〈단어 연구〉
　　transcribe 옮겨적다
　　defer 연기하다
　　ready for service 운항 가능한 상태로 되어 있는
　　consider 생각하다

〈해 석〉
　　검사의 최종 절차 : 모든 작업 카드가 완료되고 그 작업이 검사원이 승인하였는가를 확인한다. 완료되지 않은 항목은 감항성에 관계가 없는 성질의 것인지 또 이들 항목이 비행 기록부에 바르게 옮겨 기록되어 있는지, 작업이 연기되어 있는지의 여부를 점검한다. 시험 비행이 필요한지의 여부, 또 그 항공기가 운항 가능한 상태로 되어 있는지를 점검한다. 모든 최종 검사가 완료되고 항공기가 감항성이 충분하다고 생각되면 레드 태그(운항 금지 태그를 말함)를 뗀다.

2) Aircraft Skill

1 Open instrument panel access door : Remove ball-lock pin securing each latch handle at aft end of door. Actuate handles in a forward direction to unlatch the spring loaded latches and push upward on door. Rotate door to the full OPEN position.

〈단어 연구〉
　　ball-lock pin 핀의 한끝의 노치를 밀면(누르면) 다른쪽 끝에 나와있는 볼이 들어가
　　　　핀이 빠지고 반대로 노치를 눌러 끼워 넣고 노치를 떼면 볼이 빠져나와 핀이 빠지지
　　　　않게 되는 핀을 말함
　　unlatch 랫치를 풀다
　　spring loaded latch 스프링으로 원래대로 돌리는 힘이 작용하고 있는 랫치

〈해 석〉
　　계기판의 액세스 도어를 연다 : 액세스 도어 끝에 달려 있는 각 랫치의 핸들을 고정하고 있는 볼 록크 핀을 떼어낸다. 핸들을 전방으로 작동시켜 스프링 힘이 작용하고 있는 랫치를 풀고 도어를 위치로 민다. 도어를 돌려서 완전 개방 위치까지 연다.

> ② Remove access doors : Using proper screwdriver, remove screws using precautions to prevent stripping the screw head. Place all screws in bag and attach to access plate. Small plates should be secured to the airplane with 1 of 2 screws. Place large plates on the parts rack. If it is necessary to drill off screw heads to remove plates or fairings, remove screw shank from nut after plate is removed.

〈단어 연구〉

precaution 주의, 경계
strip 제거하다, 벗겨내다
parts rack 부품 선반
drill off 드릴로 제거하다
fairing 페어링
screw shank 스크류 샹크, 스크류의 나사가 나있지 않은 긴 부분

〈해 석〉

액세스 도어를 장탈한다 : 적절한 스크류 드라이버를 써서 스크류 머리가 찌그러지지 않게 주의하며 스크류를 제거한다. 스크류는 전부 봉투에 넣어 액세스 플레이트에 붙여둔다. 작은 플레이트라면 한개나 두개의 스크류로 기체에 고정해둔다. 큰 플레이트는 부품 선반에 둔다. 만약 플레이트나 페어링을 장찰하는데 스크류 머리를 드릴로 제거할 필요가 있을 때는 플레이트를 떼낸 뒤 너트에서 스크류 샹크를 제거해야 한다.

3) Hydraulic Skill

> ① Preparation for work :
> (a) Actuate speed brake handle in cockpit to raise spoilers. Install lock on speed brake handle and on one cylinder of each spoiler to prevent inadvertent lowering of the spoilers.
> (b) After ascertaining that wing flaps are clear, lower the flaps to the full down position.
> (c) To open nose gear doors, pull trip levers while supporting doors to prevent sudden opening.

제3장 Aircraft Maintenance

〈단어 연구〉
preparation 준비
inadvertent 부주의한
lowering 위로 펼쳐져 있는 스포일러가 자중에 의해 원위치로 내려오는 것
ascertain 확인하다
clear 플랩이 내려오는 방향에 작업대 등이 없고 내려도 괜찮은 것을 말함
trip lever 트립 레버, 노스 랜딩기어 격납실의 도어를 여는 레버

〈해 석〉
작업 준비
(a) 조종실에 있는 스피드 브레이크 핸들을 움직여 스포일러를 올림 위치로 한다. 다음에 스피드 브레이크 핸들과 각 스포일러의 1개의 실린더에 락크를 달아 스포일러가 불시에 닫히는 것을 방지한다.
(b) 윙 플랩의 주변에 아무것도 없음을 확인하고 플랩을 하한 끝까지 내린다.
(c) 노스 랜딩기어 도어를 열 경우 도어가 갑자기 열리지 않게 도어를 받치고 트립 레버를 당긴다.

---

❷ Measuring tire pressure : With pressure gage, check tire pressure of each nose wheel tire to 124 lbs. and each main wheel tire to 148 lbs.

〈단어 연구〉
pressure gage 타이어의 공기압을 측정하는 게이지

〈해 석〉
타이어압의 측정 : 압력 게이지로 각 노우즈 휠 타이어의 압력이 124 lb, 각 메인 휠 타이어의 압력이 148 lb가 되는가 점검한다.

---

❸ Maintenance of shock strut : Check shock strut for proper distance between center of torque arm end bolts as shown on instruction placard attached to shock strut. With clean dry rag, wipe strut piston clean and coat with a thin film of hydraulic fluid MIL-H-5606.

3-3. Service Work Items 97

〈단어 연구〉

shock strut 바퀴의 완충 지주, 쇽크 스트러트
torque arm 토큐 암, 쇽크 스트러트의 실린더와 피스톤부를 결합하고 있는 암
instruction placard 설명 또는 지시 플래카드
coat 막을 만들다
a thin film 얇은 피막

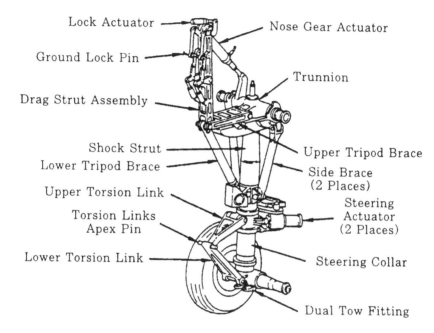

그림 3-4 Landing Gear

〈해 석〉

쇽크 스트러트의 정비 : 쇽크 스트러트의 토큐 암 끝의 볼트 중심간의 거리가 쇽크 스트러트에 달려 있는 인스트럭션 플래카드에 표시되어 있듯이 알맞는 거리로 되어 있는지 점검한다. 마른 깨끗한 천으로 스트러트의 피스톤부를 깨끗이 하고 MIL-H-5606의 작동유로 얇은 피막을 만들어주도록 한다.

④ Inspection of accumulator for leaks, condition and security : Check accumulator assembly and tubing connections for evidence of leakage and loose or broken safety wire. Remove vent plug and check for internal

제3장 Aircraft Maintenance

> leakage. If internal leakage is evident, replace accumulator on nonroutine card. If no internal leakage is found, reinstall and safety vent plug.

〈단어 연구〉
vent plug 벤트 플러그, 어큐뮬레이터의 공기를 배출시키는 밸브
evidence 흔적

〈해 석〉
어큐뮬레이터의 누출, 상태 및 부착 상태에 대한 검사 : 어큐뮬레이터와 배관의 결합부에 누출된 흔적이나 느슨함이 없는지 또는 안전선이 끊어져 있지 않은지 점검한다. 벤트 플러그를 풀고 내부 유출이 없는지 점검한다. 내부 유출이 있으면 난 루틴 카드에 따라 어큐뮬레이터를 교환한다. 내부 유출이 없으면 벤트 플러그를 재장착하고 안전선을 설치해준다.

4) Engine Skill

> ① Open nacelle doors : Release latches along bottom centerline, aft edge and forward edge of nacelle doors. Open nacelle doors and install support rods. Do not prop supports against engine.

〈단어 연구〉
prop 지탱하다

〈해 석〉
나셀 도어를 연다 : 엔진의 하부 중심선을 따라 부착되어 있는 래치와 나셀 도어의 전후단에 부착되어 있는 래치를 푼다. 나셀 도어를 열어 지지대를 설치한다. 지지는 엔진 기체에 세우지 않게 할 것.

> ② Check nacelle doors : With nacelle doors open, visually inspect the interior and exterior structure of the doors for cracks, security and general condition, noting specifically

the hinges and fittings. Inspect fire detector cable grommets at attach points for deterioration. Check loop evidence of rubbing or chafing engine, engine components or doors ; giving particular attention in the area of the aspirator valve. The loop must clear any object that can chafe or rub through the outer covering.

〈단어 연구〉

with~ open   ~을 연 채로, ~을 연 상태에서
fire detector   화재 감지기
cable   (여기서는) 불에 의해 뜨거워지면 전기적으로 경보 장치를 작동시키는 케이블을 말함
grommet   그로메트, 케이블이나 파이프 그밖의 전기 배선 등이 크램프나 구조물에 직접 닿아 손상되지 않게 삽입하는 고무 또는 플라스틱제의 완충 부품
loop   루프, 화재감지기 케이블 전체를 일컬음
rubbing   문지름 접촉
chafing   마멸(서로 마찰해서)
particular attention   특별한 주의
aspirator valve   애스피레이터 밸브, 엔진 본체와 너셀 도어 사이의 공기(연료가 샐 경우 혼합기가 되어 폭발할 위험성이 있음)를 외부로 강제적으로 배출하는 밸브
outer covering   아우터 커버링, 감지기 케이블의 외부 피복을 말함

〈해 석〉

나셀 도어의 점검 : 나셀 도어를 연 상태에서 도어의 내외면을 크랙이나 부착 상태 및 일반 상태 특히 힌지나 핏팅에 대해 자세히 검사한다. 화재 감지기 케이블의 고정 부분에 사용되고 있는 그로메트가 노화되어 있지 않은지 검사한다. 디텍터의 루프에 엔진이나 엔진 보기 또는 도어와 접속 찰상의 흔적이 없는지, 또 애스피레이터 밸브의 주변을 특히 주의해서 점검한다. 루프는 외부 피복을 찰상하거나 마멸시키는 물체에서 충분히 떨어져 있어야 한다.

3 Accessory Check : Check the following accessories for condition, security, and leakage ; C.S.D., C.S.D air oil cooler, C.S.D. air oil cooler shut-off valve, C.S.D. oil thermo-sensor, load pressure ratio transmitter, fuel flow transmitter, oil pressure switch, fuel pressure switch, air turbine starter, air turbine starter regulator, oil pressure transmitter.

제3장 Aircraft Maintenance

〈단어 연구〉
  shut-off valve 차단 밸브, 셧 오프 밸브
  thermo-sensor 온도 감지기, 온도 센서
  load pressure ratio CSD의 유압비로서 이 비가 커진다는 것은 CSD에 걸리는 하중이 증가함을 나타낸다.

〈해 석〉
  장비품의 점검 : 다음 장비품의 외관, 부착 상태 및 유출의 유무를 점검한다. CSD, CSD 작동유의 공기에 의한 냉각기, 냉각기 차단 밸브, CSD 작동유의 온도 센서, 로드 프렛셔 레시오 트랜스미터, 연료 유량 트랜스미터, 유압 스위치, 연압 스위치, 공기 터빈 스타터, 공기 터빈 스타터 조정기 및 유압 트랜스미터

④ Check engine starter oil level : Check oil quantity in engine starter. If oil is low, remove the oil fill plug on the side of the starter and fill the starter with MIL-L-7808C oil to the level of the oil fill plug. Oil will extend into the heli-coil threads of the plug port. Install the plug, tighten and lockwire.

〈단어 연구〉
  fill plug 필 플러그, 오일 보충구의 플러그
  extend 이르다, 달하다
  heli-coil thread 헬리코일을 암나사에 삽입하여 수나사가 느슨해짐을 방지하는 나사산을 말함
  lockwire 안전선을 거는 것

〈해 석〉
  엔진 시동기 오일 레벨의 점검 : 엔진 시동기의 오일량을 점검한다. 만약 오일 레벨이 낮으면 시동기 옆에 붙어 있는 오일 보급 플러그를 때내어 플러그가 있는 높이까지 MIL-L-7808C 오일을 보충한다. 오일은 보급구의 헬리코일 나사에 달할 때까지 넣는다. 플러그를 닫고 조인 후 안전선을 설치한다.

  5) Electrical Skill

3-3. Service Work Items

1 Operation Test in cockpit :
(a) Perform overheat warning test.
(b) For controller test, actuate test switch in cockpit and observe malfunction light operation. To check temperaure selector, rotate selector knob and observe temperautre readings.
(c) Turn pitot heaters on and observe that pitot heat light comes on and pitot tubes heat. Do not touch heater. Turn pitot heater off and observe that pitot tubes cool and that light goes off.
(d) Actuate each call switch and observe operation of call switch light and light on cabin attendant's control panel. Replace bulbs not burning.
(e) Check operation of door warning lights.
(f) Replace missing lamps and fuses in spare fuse box per placards.

〈단어 연구〉

malfunction 작동하지 않는, 고장나 있는
temperature reading 온도계의 지시
pitot 피토
burn (등불이) 빛나다
missing lamps 없어진 전구
spare fuse box 스페어 퓨즈 박스, 예비의 전구와 퓨즈를 넣어두는 상자
placard 플래카드, 박스 속의 어느 부분에, 어떤 규격의 전구나 퓨즈를 넣으면 좋은지를 나타내는 설명도, 박스 뚜껑 안쪽에 붙어 있다.

〈해 석〉

조종실에서의 작동 시험 :
(a) 과열 경보 장치의 시험을 하시오.
(b) 컨트롤러의 시험을 하려면 조종실의 테스트 스위치를 움직여 고장 라이트가 점등되는지를 본다. 온도의 셀렉터를 점검하려면 셀렉터 노브를 돌려 온도계의 지시치를 본다.
(c) 피토 히터를 들고 피토 히트 라이트가 점등되어 피토관이 뜨거워지는지를 확인한다. 히터를 만져서는 안된다. 피토 히터를 끄고 피토관이 식어 라이트가 꺼짐을 확인한다.

제3장 Aircraft Maintenance

(d) 각 호출 스위치를 움직여 호출 스위치 라이트와 객실 승무원용 컨트롤 패널에 있는 라이트가 점등하는지를 본다. 점등하지 않는 전구는 교환한다.
(e) 도어 경고 라이트가 점등하는지를 점검한다.
(f) 스페어 퓨즈 박스 내의 없어진 전구나 퓨즈를 플래 카드에 따라 보충한다.

---

2 Operation Check of Engine Valve :
(a) Pull fire pull handle at top of instrument panel to pull out position and note that light of Second Officer's (F/E) panel comes on to indicate valve is closed. Check the indicator on the valve in the pylon to see that it indicates "CLOSED".
(b) Check each handle micro switch for condition and security of mounting.
(c) Push fire pull handle all the way in and note the light on Second Officer's panel goes out to indicate valve is open. Check indicator on valve to see that it indicates "OPEN".
(d) Actuate shut-off valve switches and check "CLOSED" and "OPEN" positions.
Note : Check each of the engine valves one at a time.

〈단어 연구〉

engine valve 엔진 밸브. 엔진에서 고온, 고압의 공기(pneumatic air or bleed air)를 보내거나 차단하는 밸브
fire pull handle 파이어 풀 핸들. 엔진의 화재시에 당기는 핸들로서 이것을 당기면 엔진의 브리드 에어 외에 연료, 작동유 등 화재를 확대시킬 수 있는 기체, 유체를 차단하게 되어 있다.
micro switch 마이크로 스위치. 이 기체에는 엔진 밸브를 전기적으로 컨트롤하므로 핸들을 당기면 작동하는 마이크로 스위치가 달려 있다.
all the way in 최대한 밀어넣다
one at a time 1번에 1개씩

〈해 석〉

엔진 밸브의 작동 점검 :
(a) 계기판의 맨위에 있는 파이어 풀 핸들을 최대한 당겨 항공 기관사의 계기판에 있는 라이트가 점등되고 밸브가 닫힌 것을 나타내는지 확인한다. 파일론 내의 밸브에 달려

있는 지시계가「CLOSED」를 지시하는지를 보고 확인한다.
(b) 각 핸들에 있는 마이크로 스위치의 외관 및 부착 상태를 점검한다.
(c) 파이어 풀 핸들을 최대한 밀어넣어 항공 기관사의 계기판에 있는 라이트가 꺼지고 밸브가 열린 것을 나타냄을 확인한다. 밸브에 달려있는 지시계가「OPEN」를 가리키고 있는지를 보고 확인한다.
(d) 차단 밸브의 스위치를 움직여(엔진 밸브의 인디케이터가)「CLOSED」및「OPEN」위치가 되는 것을 점검한다.
주 : 각 엔진 밸브는 1번에 1개씩 점검할 것.

---

3 Check engine instrument : Check the following engine instruments, plumbing, wiring cables and connections for condition, security and leaks.
(a) Tachometer generator.
(b) Oil quantity transmitter.
(c) Low engine oil pressure warning switch.
(d) Engine oil pressure transmitter.
(e) Fuel flow transmitter.
(f) Fuel pressure warning switch(2 each).
(g) Engine pressure ratio transmitter.
(h) Hydraulic pressure warning switch(pylon area).

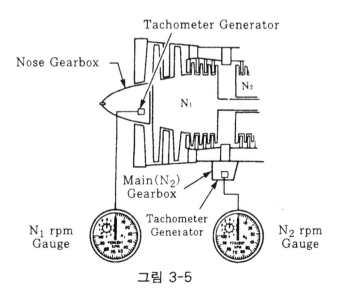

그림 3-5

〈단어 연구〉

plumbing 배관
wiring cable 전기 배선
tachometer 회전계. 2축의 젯트 엔진에서는 저압 터빈/압축기의 회전계를 N_1 타코미터, 고압 터빈/압축기의 회전계를 N_2 타코미터라고 부른다.
Engine pressure ratio 엔진 압력비. 엔진의 공기 흡입구의 압력과 배기구의 압력과의 비를 말하며, EPR이라고 생략해서 부른다.

〈해 석〉

엔진 계기의 점검 : 엔진 계기와 그 배관, 배선 및 결합부의 외관이나 부착 상태 그리고 유출을 점검한다.
(a) 회전계용 발전기
(b) 오일량 트랜스미터
(c) 엔진 오일 저압 경보 스위치
(d) 엔진 오일 압력 트랜스미터
(e) 연료 흐름 트랜스미터
(f) 연압 경보 스위치(2개)
(g) 엔진 압력비 트랜스미터
(h) 작동 유압 경보 스위치(파일론부)

3-4. Discrepancy

 항공기는 사용함으로서 그 구조나 계통 또는 기능 부품 등에 고장이나 불량 상태가 생김은 피할 수 없는 일이다. 이들 이상 상태를 발견하여 수리하는 것이 정비의 큰 목적의 하나인데, 항공기는 공중을 안전히 운항한다고 하는 다른 탈 것에선 볼 수 없는 엄격한 조건이 요구되므로 항공기의 상태를 정확히 파악하기 위한 원자료로서 또는 사고가 있을 때의 중요 자료로서 발생한 고장이나 불량 상태와 그들을 어떻게 수리했는지를 기록에 남길 필요가 있다.
 또 수리를 할 때 작업자에 대해 고장의 내용과 부위를 명확히 전하는 생산 관리상의 수단도 되기 때문에 정비에 종사하는 사람은 고장의 내용, 부위 및 그 수리 내용, 방법 등을 적절히 기록하는 것이 기본적 능력으로 요구된다. 고장이나 불량 상태를 Discrepancy(경우에 따라서는 Squawk)라고 하며 이들의 수리를 Correction이라고 하는데 다음에서 이들이 영어로 어떻게 표현되고 있는지 살펴보기로 하자.

1) Ground Discrepancies

> 1 Loose knob on No.4 fuel control lever on flight engineer's panel.
> Corr. : Tightened the screw and secured the knob.

〈단어 연구〉
　flight engineer's panel　항공 기관사가 담당하는 계기나 레버, 스위치 등이 달려 있는 조작판을 말하며, 통상 F/E panel로 생략되는 경우가 많다.
　corr.　수리, 수정(correction의 약자)

〈해 석〉
　항공 기관사 계기판의 제4번 연료 탱크 제어 레버 끝에 달려 있는 잡는 노브가 느슨해져 있다.
　수리 : 나사를 조여 노브가 움직이지 않게 했다.

제3장 Aircraft Maintenance

> **2** Loose nuts attaching tacho-generator in No.3 engine.
> Corr : Tightened the nuts.

〈단어 연구〉

　tacho-generator 엔진 회전계의 신호원이 되는 발전기.

〈해 석〉

　제3번 엔진의 타코 제너레이터를 장착하고 있는 너트가 몇 개 느슨해져 있다.
　수리 : 너트를 조였다.

> **3** Oil leakage at the connection of hydraulic pressure line at forward, inboard corner in right hand Main Landing Gear well.
> Corr : Found a B-nut loose. Tightened the nut but still leaked. Replaced the pipe and confirmed no leak with pressurized C′K.

〈단어 연구〉

　leakage 샘, 누출
　at the connection of ~의 연결 부분에서
　hydraulic pressure line 유압 계통의 압력 라인
　forward 전방, Fwd로 줄여 쓴다(↔after 후방, Aft)
　inboard 내측, Inb′d로 표기한다(↔ outboard 외측, Outb′d)
　right hand 우측, R.H로 표기한다(↔Left hand 좌측, L.H)
　Main Landing Gear well 메인 랜딩기어 격납실(M.L.G well로 표기한다)
　B-nut B 너트, 배관을 이을 경우에 쓰이는 너트
　C′K 시험(check의 약자)

〈해 석〉

　우측 메인 랜딩기어 웰의 전방 내측 코너 있는 곳에서 유압 계통의 압력 라인에서 작동유의 유출이 있다.
　수리 : B너트의 느슨함 발견. 너트를 조였으나 누출이 멈추지 않았다. 파이프를 교환하고 압력 첵크를 해서 유출이 없음을 확인했다.

3-4. Discrepancy

4 Fuel leakage at several rivets in lower wing skin(as marked) outb'd near No.1 engine.
Corr. : Forced sealant into around each rivet.

〈단어 연구〉

as marked 표시되어 있듯이〔익면과 같이 넓고 특히 표시가 되는 것이 없는 장소에서는 고장 부위를 나타내기 위해 백묵이나 잘 지워지지 않는 그래프 펜슬 등으로 표시해 두는 방법이 쓰인다. 이 경우 기체의 재료를 손상할 예리한 것이나 매직 잉크와 같이 지우기 힘든 것의 사용은 패해야 한다. 또 작업자라면 모두가 알고 있는 장소나 장비품(여기서는 1번 엔진)을 기준으로 해서 표시하는 것이 일반적인 방법이다〕
force into ~을 밀어 넣다.
sealant 실런트, 틈 등을 메꾸는 화학 재료를 말함.
주 : fuel leakage의 종류에는 다음과 같은 것이 있다.
 stain : 약간 스며나오는 정도의 누출
 seepage : 직경 5~10cm 정도로 번짐이 퍼져있는 누출
 dripping : 똑똑 떨어지는 정도의 누출
 running : 흐르는 정도의 누출
 wet : 전반적으로 젖어 있는 상태. 새는 장소가 확실하지 않는 경우가 많다.

〈해 석〉

No.1 엔진의 외측 부근의 날개 밑면 몇개의 리벳(표시가 되어 있는)에서 연료가 새고 있다.
수리 : 각 리벳의 주위에 실런트를 넣었다.

5 Nose tire worn and cut.
Corr. : Replaced nose tire.

〈단어 연구〉

Nose tire 노스기어의 타이어
worn 마모되어 있는
cut 자르다

〈해 석〉

노스기어의 타이어가 마모되고 또 잘린 흔적이 있다.
수리 : 노스기어의 타이어를 교환했다.

제3장 Aircraft Maintenance

> **6** No.4 M.L.G. brake leaks and worn out.
> Corr : Replaced brake unit and perfomed bleeding.

〈단어 연구〉
bleed 브리드하다〔유압으로 작동하는 부품을 교환한 경우는 유압 파이프 내에 남아 있는 공기를 빼낼(purge) 필요가 있다. 이것을 브리드라고 한다〕.

〈해 석〉
메인 랜딩기어의 제4번 브레이크에서 작동유가 새고 브레이크가 마모되어 있다.
수리 : 브레이크를 교환하고 브리드를 했다.

> **7** Missing screws for attaching access door L-24 on wing leading edge lower between No.3 and No.4 engines.
> Corr. : Fixed up with new screws.

〈단어 연구〉
missing 빠져 있는, 없어진
access door 액세스 도어, 정비용으로 동체나 날개면에 뚫려 있는 구멍을 막아두는 판〔엑세스 프레이트(access plate)라고도 하며 식별 번호가 붙어 있다〕
fix up 부착하다, 수리하다

〈해 석〉
주익 리딩에이지 밑면의 No.3 및 No.4 엔진 사이에 있는 L-24의 액세스 도어를 고정하고 있는 스크류가 몇개 떨어져 있다.
수리 : 새로운 스크류로 조였다.

> **8** Missing cotter pin in the lock bolt of L.H M.L.G. wheel retaining nut.
> Corr. : Installed cotter pin.

〈단어 연구〉
cotter pin 코터 핀, 너트의 느슨함을 방지하는 핀
lock bolt 락크 볼트, 느슨함을 막는 볼트, 고정 볼트

3-4. Discrepancy

retaining nut 어떤 부품을 유지하는 너트

〈해 석〉

좌측 메인 랜딩기어 휠의 리테이닝 너트 고정 볼트의 코터 핀이 없어졌다.
수리 : 코터 핀을 장착했다.

---

9 Rusty heads of screws in upper skin of L.H wing tip.
Corr. : Polished the heads of screws with sand paper and applied cat-a-lac paint.

〈단어 연구〉

rusty 녹슬어 있는
polish 갈다, 닦다
sandpaper 샌드 페이퍼〔천 위에 금강사를 부착시킨 것을 에메리천(Emery cloth)이라고 한다〕
cat-a-lac 캣타랙, 내산성 및 내스카이드롤성의 페인트명

〈해 석〉

좌측 윙 팁 상면의 스크류 머리가 몇개 녹슬어 있다.
수리 : 샌드 페이퍼로 스크류 머리를 갈고 캣터랙 페인트를 칠했다.

---

10 1 inch crack in L.H wing trailing edge upper skin at wing station number 120.
Corr. : Drilled stopping holes at both ends of the crack.

〈단어 연구〉

wing station number 윙 스테이션 넘머, 날개의 각부의 위치를 나타내기 위해 동체의 중심에서 윙 팁(익단) 방향으로 1 in 마다 표시한 치수를 말함
stopping hole 균열의 끝에 뚫는 작은 구멍으로 균열이 그 이상 진행되지 않게 하는 처치

〈해 석〉

우측 트레일링에이지 윗면의 윙 스테이션 120에 1 in의 균열이 있다.
수리 : 균열의 양끝에 스톱 홀을 뚫었다.

제3장 Aircraft Maintenance

> ⑪ 3 inch crack in fuselage bottom skin at fuselage station number 650 at 7 o'clock (looking forward).
> Corr. : Fixed with plug patch.

〈단어 연구〉
  fuselage bottom 동체 밑면
  fuselage station number 동체 스테이션 넘버. 동체 각부의 위치를 나타내기 위해 동체 앞 기준선에서부터 후방으로 1 in 간격으로 정해진다.
  looking forward 후방에서 앞쪽을 보아
  plug patch 플러그 패치. 외부의 수리 방법의 하나로 손상 부분을 잘라내고 안에서 판을 댐과 동시에 떼어낸 형과 같은 형의 판을 끼워 넣어 리벳으로 부착하여 외피와 표면이 평평하게 되도록 하는 수리 방법이다.

〈해 석〉
  동체 밑면의 스테이션 650의 7시의 위치(뒤에서 보아)에 3 in의 균열이 있다.
  수리 : 플러그 패치로 수리했다.

> ⑫ Electrical wire conduit and aileron control cable are chafing in L.H flap well at wing station 310.
> Corr. : Re-routed the conduit and kept it off the cable.

〈단어 연구〉
  conduit 컨디트. 전선을 보호하기 위해 씌워놓은 알루미늄 파이프
  chafe 문지르다
  flap well 플랩 웰. 플랩의 리딩에이지 부분이 플랩을 올린 위치에서 격납되는 주익 트레일링에이지에 설치된 공동 부분으로서, 이 속에 케이블, 파이프, 덕트 및 와이어 등이 뻗어 있다.
  re-route (전기 배선)을 기존의 위치로부터 변경시켜 재배선한다

〈해 석〉
  전기 배선의 콘디트와 보조 날개의 작동 케이블이 좌측 플랩 웰 내 주익 스테이션 310 부근에서 접촉하여 닳아 있다.
  수리 : 콘디트의 위치를 재배치하여 케이블에서 멀어지게 했다.

⑬ Dent (2″×1″) on No.3 engine inlet cowl tip at 4 o'clock (looking after).
Corr. : OK for service.

〈단어 연구〉

dent 덴트, 패임, 눌린 자국
engine inlet cowl tip 엔진 공기 흡입구 선단
looking after 전방에서 후방으로 보아
OK for service 운항에 지장 없는, 불량 장소가 있어도 운항에 지장이 없다고 판단했을 때는 특히 수리를 하지 않는 경우가 있다.

〈해 석〉

No.3 엔진 공기 흡입구 가장자리 4시 방향에 가로 2 in, 세로 1in의 패임이 있다. (후방으로 보아서)
수리 : 운항에 지장이 없으므로 특별히 수리하지 않았다.

⑭ Paint peeled off upper side of radome.
Corr. : Applied paint

〈단어 연구〉

peel off 벗겨져 있는
radome 레이돔, 기수 선단에 탑재되어 있는 기상용 레이더를 덮고 있는 돔을 말함

〈해 석〉

레이돔 상면의 페인트가 벗겨져 있다.
수리 : 페인트를 칠했다.

⑮ Aileron control operates "stiff" over the whole travel.
Corr. : Found cable off pulley at R.H wing root leading edge. Replaced the cable on pulley and corrected the cable tension.

112 제3장 Aircraft Maintenance

〈단어 연구〉
 stiff 단단한, 굳은
 travel 행정
 replace 원래 위치로 돌리다, 교환하다
 cable tension 케이블 텐션, 케이블의 장력(케이블의 크기로 표준값이 정해지며, 온도에 따라 보정한다. 이것을 측정하는 기구를 tension meter 라고 한다)

〈해 석〉
 에일러론 컨트롤이 전행정에 걸쳐 움직임이 빡빡하다.
 수리: 우측 주익 리딩에이지 루트에서 케이블이 풀리에서 벗어나 있다. 케이블을 원래대로 걸어 케이블 텐션을 조절했다.

⑯ Elevator control operates "jerkey" toward up position.
Corr. : Found control cable frayed and catching on pulley in tail compartment. Replaced the defective cable and corrected cable tension.

〈단어 연구〉
 jerkey 걸린
 fray 쏠려 끊겨져 있는

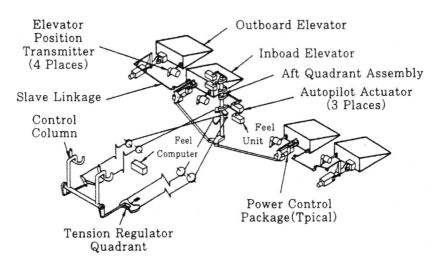

그림 3-6

catching 걸려 있는
defective 손상되어 있는

〈해 석〉

엘리베이터 컨트롤이 올린 방향에서 움직임에 걸림이 있다.
수리 : 컨트롤 케이블이 쓸려 끊어져 테일 컴파트먼트 내의 풀리에 걸려 있었다. 손상된 케이블을 교환하고 케이블 텐션을 조절했다.

⑰ Flaps "up" visually, but indicator reads "down".
Corr. : Replaced the indicator. Operational C′K OK.

〈해 석〉

플랩이 올린 위치로 되어 있는데도 계기는 내린 위치를 나타내고 있다.
수리 : 계기를 교환했다. 작동 시험을 해서 정상임을 확인했다.

⑱ Taxi light on nose landing gear inoperative, and the C.B won't stay at stowed position.
Corr. : Replaced the C.B Operationally checked OK.

〈단어 연구〉

 taxi light 택시 라이트. 야간 지상을 주행할 경우 전방의 지상을 비추는 라이트
 inoperative 작동하지 않는 여기서는 점등하지 않는다는 의미. C.B 서킷 브레이커
   (circuit breaker의 약자)
 stowed position 밀어 넣은 위치. 즉 전류가 흐르는 위치

〈해 석〉

노스 랜딩기어에 달려 있는 택시 라이트가 점등하지 않는다. 또 서킷 브레이커가 들어간 위치에서 유지되지 않는다.
수리 : 서킷 브레이커를 교환한 후 작동 점검을 해서 정상이다.

⑲ L.H lamp of landing light fails to light.
Corr. : Found loose connection just behind control SW. Tighten the connection. Operational C′K OK.

〈해 석〉
랜딩 라이트의 좌측이 점등하지 않는다.
수리 : 컨트롤 스위치의 안쪽의 접속 부분이 느슨해져 있다. 접속 부분을 조였다. 작동 시험을 해서 점등함을 확인했다.

---

⑳ No.1 engine was failure to start.
Corr. : Found spark plugs wet and high tension lead connectors dirty.
Cleaned up wet spark plugs and the high tension lead connectors.

〈해 석〉
No.1 엔진이 시동되지 않는다.
수리 : 점화 플러그가 젖어 있고 하이 텐션 리드의 접속 부분이 더러워져 있다. 점화 플러그 및 하이 텐션 리드의 접속 부분을 깨끗이 했다.

---

⑳ No.3 engine oil pressure low.
Corr. : Replaced the oil pressure indicator. Engine run up check OK.

〈해 석〉
No.3 엔진의 오일 압력이 낮다.
수리 : 오일 압력계를 교환했다. 엔진 작동 시험으로 정상이 된 것을 확인했다.

---

⑳ No.1 engine propeller failure to feather.
Corr. : Generator control does not provide enough travel to allow governor to move hard against high pitch stop. Adjusted to provide sufficient travel in governor control.

〈해 석〉
No.1 엔진의 프로펠러가 페더링 상태로 되지 않는다.
수리 : 거버너 컨트롤이 거버너를 하이 피치 스톱에 세게 닿도록할 충분한 행정을 갖고 있지 않다. 거버너 컨트롤에 충분한 행정이 가능하도록 조정했다.

3-4. Discrepancy   115

> ⑳ No reading on No.2 engine tachometer indicator.
> Corr. : Replaced the tachometer generator as found no output.

〈해 석〉

No.2 엔진의 회전계가 지시를 하지 않는다.
수리 : 회전계용 발전기의 출력이 나오지 않아서 발전기를 교환했다.

2) Flight Discrepancies

> ① Felt insufficient braking action while taxiing.
> Corr. : Found leakage at L.H shuttle valve and brake lining worn out. Replaced brake assembly and tightened the connections of hydraulic lines at shuttle valve. Performed bleeding the brake system and leakage check OK.

〈단어 연구〉

　shuttle valve 셔틀 밸브. 작동유의 유압이 없어져 비상용의 압축 공기를 사용할 경우, 자동적으로 절환되는 밸브
　brake lining 브레이크 라이닝. 브레이크의 회전판에 압접하여 브레이크를 거는 고정판에 부착되어 있는 재료를 말함

〈해 석〉

　지상 주행중 브레이크의 성능이 충분하지 않았다.
　수리 : 좌측의 셔틀 밸브에서 유출되어 브레이크 라이닝이 마모되어 있었다. 브레이크 어셈블리를 교환하고 셔틀 밸브의 유압 파이프의 접합부를 조였다. 브레이크 계통을 브리드하여 유출 시험으로 새지 않음을 확인했다.

> ② Airspeed indicator hand oscilates at faster speed than 250 mph.
> Corr. : Found leakage in pitot line. Corrected the leakage. Confirmed the system operated normally by test hop.

제3장 Aircraft Maintenance

〈단어 연구〉

oscilate 진동하다
mph 시간당 마일(mile per hour : 속도의 마일 단어)
test hop 시험 비행

〈해 석〉

대기 속도계의 지침이 시속 250mile 이상의 속도에서 진동한다.
수리 : 피토 배관에 유출이 발견 되었다. 새는 것을 고쳤다. 시험 비행을 해서 피토 시스템이 정상으로 작동함을 확인했다.

---

3 Landing gear unsafe light came "ON" during climb(25 ft, 300kt). Made up-latch check and cycled gear operation at 230kt but the light did't go out.
Corr. : All landing gear up limit switches' function C'K OK by tester. Adjusted R.H M.L.G door actuator rod about half turn to shorten. Confirmed gear operation and indication all normal on jacks.

〈단어 연구〉

gear unsafe light 기어 언세이프 라이트, 바퀴가 올라간 위치에 확실히 고정되어 있지 않을 때 점등하는 적색의 경보등을 말함.
kt 노트(knot의 약자)
up-latch check 올린 위치에서 래치가 걸려져 있는지 점검하는 것
cycle gear operation 기어의 올리고 내림을 반복하는 것
shorten 짧게 하다

〈해 석〉

착륙 장치의 언세이프 라이트가 상승중 25,000ft, 300kt에서 점등했다. 업 래치 첵크를 해서 230kt에서 기어를 올리고 내림을 반복했으나 언세이프 라이트가 꺼지지 않았다.
수리 : 기어 업 리미트 스위치 기능을 테스터로 조사했으나 전부 정상이었다. 우측 메인 랜딩기어 도어의 작동 로드를 반회전 돌려서 짧아지게 조절했다. 잭키 업을 하고 랜딩 기어의 작동과 그 지시가 전부 정상임을 확인했다.

3-5. Modification or Alteration

 항공기의 수리 또는 개조는 제조자가 발행하는 서비스 불러틴이나 그 내용 중 특정한 것에 대해 항공국이 실시를 의무화한 감항성 개선 명령에 따른 경우가 많다.
 미국연방항공국(Federal Aviation Administration 약하여 FAA라고 함)이 명령으로 낸 것을 Airworthiness Directives (감항성 개선 명령) 흔히 AD라고 한다.

1) Service Bulletin

> **1** Nose Landing Gear :
> (a) Landing gear ; Nose wheel steering cylinder assemblies — Rework of elbow orifice valve assemblies.
> (b) Reason : Operators have reported instances where particles from deteriorated "O" rings and Teflon back up rings in the nose wheel steering cylinders have entered the elbow orifice valve. Investigation indicates that improper maintenance or towing the aircraft without disconnecting the torque links can cause excessive pressures and packing damage. To reduce the possibility of contaminates entering and clogging the elbow orifice check valves, the valves have been revised by the incorporation of a filter feature.
> (c) Compliance ; Although the work outlined herein does not affect the immediate safety of the aircraft, the DAC considers it highly desirable and should be accomplished at the earliest practicable date.

〈단어 연구〉
 rework 작업(재작업)
 orifice 작은 구멍, 오리피스

제3장 Aircraft Maintenance

operator 오퍼레이터, 항공기의 사용자, 항공 회사
instance 사례
particle 분자, 작은 조각
"O" ring 오링, 덕트나 파이프의 결합 부분에 사용되는 유출 방지 시일
Teflon back up ring 테프론 백 업 링, 실린더와 피스톤의 습동으로 당겨져 못쓰게 될 염려가 있으므로 오링과 함께 비교적 굳은 테프론제의 링을 삽입하여 오링이 본래의 기능을 발휘할 수 있게 보조하는 링
investigation 조사
improper 부적당한
towing 토잉, 견인차로 끌거나 밀어서 항공기를 이동시키는 것
torque link 토큐 링크, 랜딩기어 스트러트의 실린더부와 피스톤부를 결합하고 있는 연결간
packing 팩킹.
contaminate 오염시키다. (복수로) 오염 물질
clogging 막힘
revise 수정하다
incorporate 짜넣다
feature 특질, 특색
compliance 상용의 완급도
the DAC 더글러스사를 말함(the Douglas Aircraft Company의 약자)
practicable 실시 가능한

〈해 석〉

노스 랜딩기어 :
(a) 랜딩기어 : 노스 랜딩기어 스티어링 실린더 어셈블리 — 엘보우 오리피스 밸브 어셈블리의 수리
(b) 이유 : 항공 회사가 노스 랜딩기어 스티어링 실린더에 사용하고 있는 오링과 테프론 백업 링이 노화되어 그 작은 조각이 엘보우 오리피스 밸브에 들어간 사례가 경고되었다. 조사한 바에 의하면 부적절한 정비 행위 및 토큐 링크의 결합을 풀지 않은 채 항공기를 견인하면 과도한 압력을 유발시켜 팩킹을 손상시킬 수 있다고 한다. 엘보우 오리피스 첵크 밸브에 더러움이 생겨 막힐 가능성을 적게 하기 위해 밸브에 필터를 넣어 수정했다.
(c) 적용의 완급 : 여기서 설명하는 수리는 항공기의 안전에 직접 영향을 주지 않지만 더글러스사는 이 수리가 매우 바람직하다고 생각하여 가장 빠른 시기에 실시해야 한다고 여기고 있다.

2) Airworthiness Directives

3-5. Modification or Alteration

> **1** Beech. Amendment 39-1371
> (a) Compliance ; Required as indicated unless already accomplished. To prevent engine damage from loss of lubricant due to accidental opening of oil sump drains by age stiffened plastic drain hoses, within 25 hours time in service after the effective date of this AD, accomplish the following.
> (b) Remove the oil drain tubes from the sump drain valves. Beechcraft Service Instructions No.0473241 pertains to this same subject. This amendment becomes effective January 5, 1992.

〈단어 연구〉

amendment 수정, 개정
unless ~가 아니면
due to ~가 원인으로
accidental 우발적인
age stiffened 세월이 흘러 경화
time in service 비행 시간
pertain to ~에 관한, 관련된

〈해 석〉

비치사 개정 39-1371 호
(a) 적용의 완급도 : 이미 실시되고 있는 경우를 제외하고 여기 설명되어 있듯이 실시해야 한다. 플라스틱의 드레인 호스가 노후 경화됨에 따라 오일 섬프 드레인이 우발적으로 열려 오일이 없어져 엔진이 손상되는 것을 방지하기 위해 이 개선 명령의 **발효후**, 비행 시간이 25시간 이내에 다음 사항을 실시한다.
(b) 섬프 드레인 밸브에서 드레인 튜브를 제거한다. 비치 크래프트 서비스 인스트럭션 0473241호가 이것과 같은 주제의 것이다. 이 개정은 1992년 1월 5일부터 **효력을** 발한다.

> **2** Boeing. Amendment 39-1379
> (a) Compliance required as indicated. To detect cracks in the main landing gear downlock torque shaft, accomplish the following.

120　제3장 Aircraft Maintenance

AIRWORTHINESS DIRECTIVE

AVIATION STANDARDS NATIONAL FIELD OFFICE
P.O. BOX 26460
OKLAHOMA CITY, OKLAHOMA 73125

The following Airworthiness Directive issued by the Federal Aviation Administration in accordance with the provisions of Federal Aviation Regulations, Part 39, applies to an aircraft model of which our records indicate you may be the registered owner. Airworthiness Directives affect aviation safety. There are regulations which require compliance with them. If you are concerned that our records may operate on contracts to which an Airworthiness Directive applies, except in accordance with the requirements of the Airworthiness Directive (FAR 39.3).

91-06-17 BOEING: Amendment 39-6931. Docket No. 90-NM-164-AD.
　　Applicability: Model 747-400 series airplanes through line number 791, certificated in any category.
　　Compliance: Required as indicated, unless previously accomplished.
　　To prevent the failure of either pilot's electronic flight instrument system (EFIS) control panel, with resultant smoke in the flight deck, accomplish the following:
　　A. Within 90 days after the effective date of this AD, inspect both EFIS control panels installed on the airplane and record the serial number and modification status. Control Panels with serial numbers identified in Collins Service Bulletin DCP-7000-31-04, dated December 1, 1989, and which do not have Modification 4 implemented, must be removed and modified in accordance with the service bulletin before further flight.
　　5. Any EFIS control panel with a serial number identified in Collins Service Bulletin DCP-7000-31-04, dated December 1, 1989, and which does not have Modification 4 implemented, must be modified in accordance with the service bulletin before installation on an airplane.
　　C. An alternate means of compliance or adjustment of the compliance time, which provides an acceptable level of safety, may be used when approved by the Manager, Seattle Aircraft Certification Office (ACO), FAA, Transport Airplane Directorate.
　　NOTE: The request should be submitted directly to the Manager, Seattle ACO, and a copy sent to the cognizant FAA Principal Inspector (PI). The PI will then forward comments or concurrence to the Seattle ACO.
　　D. Special flight permits may be issued in accordance with FAR 21.197 and 21.199 to operate airplanes to a base in order to comply with the requirements of this AD.
　　All persons affected by this directive who have not already received the appropriate service documents from the manufacturer may obtain copies upon request to Boeing Commercial Airplane Group, P.O. Box 3707, Seattle, Washington 98124; and Collins Air Transport Division/Rockwell International, 400 Collins Road N.E., Cedar Rapids, Iowa 52406. These documents may be examined at the FAA, Northwest Mountain Region, Transport Airplane Directorate, 1601 Lind Avenue S.W., Renton, Washington.

2 91-06-17

　　This amendment (39-6931, AD 91-06-17) becomes effective on April 15, 1991.

FOR FURTHER INFORMATION CONTACT:

Mr. Kenneth J. Schroer, Seattle Aircraft Certification Office, Systems and Equipment Branch, ANM-130S; telephone (206) 227-2795. Mailing address: FAA, Northwest Mountain Region, Transport Airplane Directorate, 1601 Lind Avenue S.W., Renton, Washington 98055-4056.

그림 3-7 Airworthiness Directive

3-5. Modification or Alteration

(b) For all torque shafts which have accumulated 8,000 or more landing cycles on the effective date of Amendment 39-1379, inspect the shaft within the next 300 landings after the effective date of Amendment 39-1379, unless already accomplished within the last 100 landings, and thereafter at intervals not to exceed 400 landings since the last inspection, per (a) below, untill the shaft is replaced per (b) below.

〈단어 연구〉

downlock torque shaft 다운 락크 토큐 샤프트. 바퀴를 내린 위치에 고정하는 락크를 작동시키는 토큐 축을 말함.
landing cycle 착륙 회수

〈해 석〉

보잉사 개정 39-1379 호
(a) 적용의 완급도는 기술된 대로이다. 메인 랜딩기어의 다운 락크 토큐 샤프트에 발생한 크랙을 발견하려면 다음 사항을 실시한다.
(b) 개정 39-1379 호의 발효 날짜의 시점에서 8,000회 이상의 착륙 회수를 경험한 모든 토큐 샤프트에 대해 개정 39-1379호의 발행 일자 전의 착륙 회수 100회 이내에 이미 검사가 실시되어 있지 않으면 발효 일자 후의 착륙 회수 300회 이내에 그 샤프트를 검사한다. 그 뒤에는 다음 (b)항에 따라 샤프트를 교환할 때까지 하기 (a)항에 따라서 전회의 검사에서 400 착륙 회수를 넘지 않는 간격으로 반복 실시한다.

122   제3장 Aircraft Maintenance

3-6. Inspection System of FAA Repair Station

항공기는 공중을 비행하는 기계라는 점에서 특히 안전성이 많이 요구되는데 이 안전성을 확보하기 위해 정비시의 검사 기능이 매우 큰 역할을 차지한다.
여기서는 항공기의 안전성과 검사의 관계를 알기 위해 미국연방항공규정 (Federal Aviation Regulation, 약자로 FAR로 부른다)에 따른 Repair Station의 검사 규칙을 살펴보기로 한다.

1) General Manager, Inspection Department

1️⃣ He is responsible for directing, planning, and laying out details of inspection standards, methods, and procedures used by the repair station in complying with all applicable FAR's and manufacturer's recommendations.

〈단어 연구〉
　responsible for   ~에 대한 책임이 있는
　directing   명령 또는 감독하는 것
　inspection standard   검사 기준
　method   방법
　in complying with   ~를 바탕으로
　applicable   적용할 수 있는
　recommendation   권고, 추천하는 것

〈해 석〉
　검사 부장은 감독, 계획 및 적용되는 모든 연방 항공 규칙 및 제조 회사가 추천하는 사항을 바탕으로 리페어 스테이션에서 사용되는 상세한 검사 기준, 방법 및 순서의 결정에 대한 책임이 있다.

2️⃣ It will be his duty to assist, supervise and direct all personnel assigned to the Inspection Dept. It will be his duty to ascertain that all inspections are properly performed

3-6. Inspection System of FAA Repair Station

on all completed work before it is released to the public, and that the proper inspection records, reports, and forms used by this repair station are properly executed. It will also be his duty to collect and maintain a file of repair orders and inspection forms in such a manner that the specific file pertaining to an assembly or unit can be located within a reasonable length of time.

〈단어 연구〉

supervise 감독하다
asigned to 지정된
ascertain 확인하다
before it is released to the public 사용되기 전에
execute 수행하다
repair order 수리 명령서
in such a manner ~와 같은 방법으로
pertaining to ~에 관한

〈해 석〉

검사 부장의 의무는 검사부에 배속된 모든 부원에 대해 지원, 감독 및 명령을 하는 것이다. 또 모든 완성된 일에 대해 그것이 사용되기 전에 모든 검사가 적절히 행해져 있는지 이 리페어 스테이션에서 사용한 검사 기록, 보고서, 양식 등이 바르게 기입되어 있는지를 확인하는 것이다. 그리고 어셈블리나 유니트에 관한 특정한 파일이 적당한 시간 내에 찾아낼 수 있는 방법으로 수리 명령서나 검사에 사용하는 양식의 파일을 수집하거나 보관하거나 하는 것도 검사 부장의 의무이다.

3 It will be his duty to maintain a current file of pertinent FAA approved specifications and Airworthiness Directives. It will be his duty to secure and maintain technical data on all units overhauled or repaired by the station. This material will consist of manufacturers' overhaul manuals, service bulletins, related FAR and Advisory Circulars.

제3장 Aircraft Maintenance

〈단어 연구〉
current 현재 유효한, 최근의
pertinent 적절한, 관계되는
material 자료
Advisory Circular 비규칙 사항을 정한 것으로 일반적으로의 지도, 주지를 목적으로 하여 발행하는 것

〈해 석〉
검사 부장은 연방 항공국이 인가한 관계 규정 및 감항성 개선 명령으로 현재 유효한 것의 파일을 보관할 의무가 있다. 또 이 리페어 스테이션에서 오버홀이나 수리한 모든 유니트에 관한 기술상의 데이터를 보관해둘 의무가 있다. 이들 자료는 제조 회사가 발행한 오버홀 매뉴얼, 서비스 블레틴, 연방항공규정이나 어드바이저리 서큘러의 관계 부분으로 된 것이다.

> **4** He is responsible for making periodic checks on all inspection tools and the calibration of precision test equipment, and to devise and maintain a system of keeping records of checks and calibration of inspection tools and precision test equipment, taking steps to insure that the established check periods are not exceeded.

〈단어 연구〉
periodic check 정기 점검
calibration 눈금의 검정
precision test equipment 정밀 시험 기기
devise 연구하다, 고안하다
take steps 수단을 마련하다
insure 보증하다
established 설정된
exceed 넘다

〈해 석〉
검사 부장은 모든 검사 공구에 대해 정기 점검을 하며 정밀 시험 기기의 눈금을 검정할 의무가 있고 또 설정된 점검 기간을 넘는 일이 없는 수단을 마련하며 검사 공구나 정밀 시험 기기의 정기 점검 및 눈금의 검정의 실시 기록이 보존될 체계를 연구 유지해야 한다.

3-6. Inspection System of FAA Repair Station

> **5** It will be his responsibility to insure that no defective or unairworthy parts are installed in any component or, unit, and aircraft released by the repair station.

〈단어 연구〉
responsibility 임무
defective 결함이 있는
unairworthy 감항성이 없는
release 수리 또는 오버홀한 항공기나 부품에 대해 감항성을 보증하여 이의 합격을 서명함으로써 수락하는 것

〈해 석〉
검사 부장은 결함이 있거나 또는 감항성이 없는 부품이 구성품이나 장비품에 장착되지는 않았는지, 또한 항공기가 리페어 스테이션으로부터 최종 합격 판단이 되었는지를 확인할 책임이 있다.

> **6** He is responsible for the final acceptance of all incoming material including new parts, supplies, and the airworthiness of articles on which work has been performed under contract. He is also responsible for Preliminary, Hidden Damage, Inspection Continuity and Final Inspections of all items precessed by the repair station.

〈단어 연구〉
final acceptance 최종 수락
supply 보급품
under contract 계약 아래
preliminary 예비의, 사전의
hidden 숨겨진, 잠재적인
continuity 계속성
processed 처리를 한

〈해 석〉
검사 부장은 신품의 부품이나 보급품을 포함하는 모든 입하 자료의 최종 수령과 계약 아래 작업을 시킨 부품의 감항성에 대한 책임이 있다. 검사 부장은 또 리페어 스테이션

제3장 Aircraft Maintenance

에서 처리되는 모든 품목에 대한 예비 검사, 잠재 손상 검사 및 최종 검사를 실시할 책임이 있다.

2) Inspection Personnel

> 1️⃣ Inspection personnel must be throughly familiar with all inspection methods, techniques and equipment in their speciality to determine the airworthiness of an article undergoing maintenance or alteration. They must also maintain proficiency in the use of inspection aids.

〈단어 연구〉
 familiar with 정통해 있는
 technique 기술
 in their speciality 전문 분야의
 undergo ~을 받다, ~을 경험하다
 proficiency 숙달
 inspection aid 검사 기기

〈해 석〉
검사 종사자는 정비 또는 개조 품목의 감항성의 유무를 결정하기 위해 전문 분야의 모든 검사 방법, 기술 및 기기에 정통해야 한다. 검사 종사자는 또 검사 기기의 사용법에 대해 항상 숙달되어 있어야 한다.

> 2️⃣ Inspection personnel must also have available and be familiar with current specifications involving inspection tolerances, limits and procedures as set forth by manufacturer of the product undergoing inspection, and other sources of applicable inspection information such FAA Airworthiness Directive, Service Bulletins, etc⋯.

〈단어 연구〉

3-6. Inspection System of FAA Repair Station

available 유효한, 쓸모있는
specifications ~에 계속되는
involve 말아넣다, 포함하다
tolerance 허용 한도, 여유(크기의 차의)
set forth 발행하다
product 제품
source 근원

〈해 석〉

검사 종사자는 또 검사를 받는 제품의 제조 회사에서 발행된 검사의 허용 한계 및 절차를 포함한 유효한 그리고 최신의 규격을 확보하고 또 정통해야 한다. FAA의 감항성 개선 명령이나 서비스 블레틴 등과 같이 적용할 수 있는 검사 외의 정보원에 대해서도 정통해야 한다.

3) Record of Work and Inspection

> 1 Record of work : A copy of the company work order will be filed for all work accomplished. This work order will contain the signature of the mechanic and the authorized inspector who approved each unit for return to service. An appropriate work sheet or cards will be used for the overhaul of components. Parts which have been replaced will be listed and these records will be maintained by this repair station.

〈단어 연구〉

signature 서명
authorized inspector 당국에서 인정하는 검사원
return to service 항공기나 부품을 항공용으로 제공하다(투입하다)

〈해 석〉

작업 기록 : 모든 완료된 작업에 대해서는 각각 회사 작업 지시서의 사본을 파일해 두어야 한다. 이 작업 지시서에는 각 장비품이 사용할 수 있는 것이라고 승인하는 작업자와 검사원의 서명이 있어야 한다. 장비품의 오버홀에는 적절한 작업 시트나 카드가 사용되어야만 한다. 교환된 부품은 목록을 작성하고, 또 이러한 기록은 리페어 스테이션에서 보관하고 있어야 한다.

128   제3장 Aircraft Maintenance

> **2** Record of inspection : Each inspection which is being conducted, i.e. preliminary, hidden damage, continuity, periodic, dimensions, condition and any other appropriate information obtained during various stages of these inspections shall be recorded in the appropriate space on the work order or company inspection form which will be made a part of the record file.

〈단어 연구〉
  conduct 하다, 처리하다.
  i.e (라틴어) id est(=that is). 즉, 바꿔 말하면
  stage 공정, 단계

〈해 석〉
  검사 기록 : 예비 검사, 잠재 손상 검사, 계속 검사, 정기 검사, 치수 검사, 상태 검사 등의 검사중인 사항과 이들 검사의 여러 단계에서 얻어지는 기타 적절한 정보는 작업 지시서의 적절한 공간이나 회사 검사 양식에 기록하여야 하고, 이 기록된 사항은 기록 파일의 일부가 되도록 한다.

4) Control of Precision Measurement Equipment and Measurement Standard

> **1** Purpose : Precision and accuracy of measurement equipment necessary to certify airworthiness of aircraft and its components are maintained in accordance with the company's measuring control regulations and traceable to the national standards.

〈단어 연구〉
  accuracy 정확도
  certify 증명하다
  traceable 추적할 수 있는
  national standard 그 나라의 기준

3-6. Inspection System of FAA Repair Station

〈해 석〉

목적 : 항공기와 장비품의 감항성을 증명하는데 필요한 측정 기기의 정밀도와 정확성은 회사의 측정 관리 규정에 따라 유지되며, 해당 국가의 표준을 따른다.

5) Malfunction or Defect Report

[1] If a defect or malfunction is found that could result in imminent hazard to flight, the repair station will use the most expeditious method at hand to inform the FAA. In any event, the repair station will file a report, within 72 hours after the defect or malfunction is found. Such report will completely describe the nature of the defect or malfunction.

〈단어 연구〉

imminent 급박한, 긴급한
hazard 위험
expeditious 신속한
at hand 근처에 있는
in any event 어쨌든
file 제출하다

〈해 석〉

만약 비행에 긴급한 위험을 초래할 결함이나 고장이 발견되었다면 리페어 스테이션은 취할 수 있는 신속한 방법으로 연방항공국(FAA)에 통보해야 한다. 모든 사건에 있어서 리페어 스테이션은 그 결함이나 고장이 발견된 후 72시간 이내에 보고서를 제출해야 한다. 이런 리포트는 그 결함이나 고장의 성격에 대히 빠짐없이 기술해야 한다.

6) Type of Inspection

[1] Preliminary Inspection : All items to undergo maintenance will be given a preliminary inspection, upon receipt, to determine the state of preservation(functional test if appropriate) and to note any obvious defects. The results

will be noted on the company work order, one copy of which will remain with the unit undergoing maintenance until it is returned to service.

〈단어 연구〉
upon receipt 수령시에
preservation 보관, 보존
obvious 명백한, 판연한

〈해 석〉
예비 검사 : 정비를 받을 모든 품목은 수령시에 예비 검사를 실시하여 보존 상태(필요하면 기능 시험을 함)를 판정하고 분명한 결함이 있으면 그것을 기록한다. 그 결과는 회사의 작업 명령서에 기입하고 그 카피의 1장은 검사를 받는 부품이 사용될 때까지 그 유니트에 부착해두어야 한다.

❷ Hidden Damage Inspection : Prior to the commencement of any work, all units or components that have been involved in an accident will be given a thorough inspection for hidden damage. This inspection will include areas adjacent to or likely to have been affected by the obviously damaged members or components. The results of this inspection will be recorded on the company work record.

〈단어 연구〉
commencement 개시
adjacent to ~부근의
likely ~과 같은, 할 것같은

〈해 석〉
잠재 손상 검사 : 사고에 관련되었던 유니트 또는 콤퍼넌트는 작업 개시 전에 모두 잠재적인 손상이 없는지 면밀히 검사한다. 이 검사는 분명히 파손되어 있는 부재 또는 구성 부품의 근방이나 그것에 의해 영향을 받았다고 생각되는 부분에 대해 행해진다. 이 검사의 결과는 회사의 작업 기록에 기입되어야 한다.

3-6. Inspection System of FAA Repair Station

> ③ Final Inspection : This inspection will be performed, and the final airworthiness determination will be made and recorded, by an authorized inspector making such determination. These records will be retained by the repair station as required in accordance with FAR 145.

〈해 석〉

최종 검사 : 이 검사는 감항성 판정을 하는 권한있는 검사원이 수행하고 감항성의 최종 판정을 하고 기록하여야 한다. 이들 기록은 필요시 연방 항공 규칙 145조에 따라 리페어 스테이션에서 보관한다.

7) Inspection of Critical Items

> ① FAR require the designation of items of maintenance and alteration which must be inspected(required inspection items), which must include at least those items of maintenance, and alteration which could result in a failure, malfunction, or defect endangering the safe operation of the airplane if not performed properly or if improper parts are used.

〈단어 연구〉

　designation 지정, 선정
　required inspection item 검사 항목(R.I.I로 생략된다)
　endanger 위험에 빠지게 하다

〈해 석〉

연방 항공 규정은 정비나 개조시에 검사하지 않으면 안될 항목(검사 요목)의 지정을 요구하고 있다. 그들 요구는 적어도 정비나 개조시에 바르게 행해지지 않았거나 또는 부적절한 부품이 사용되었을 경우 비행기의 안전한 운항을 위험에 **빠뜨릴** 고장 또는 불량 상태 및 결함 등의 결과를 초래할 것같은 요목이 포함되어 있어야 한다.

132 제3장 Aircraft Maintenance

> **2** The regulation requires the selection and designation of specific items of maintenance which, if not accomplished properly, could result in flight failure having catastrophic result.

〈단어 연구〉
catastrophic 파괴적인

〈해 석〉
이 규정은 만약 적절히 실시되지 않으면 비행중에 파손되어 파멸적인 결과를 야기시킬 수 있는 특별한 정비 항목의 선정 및 지정을 요구하고 있다.

8) Non-destructive Inspection(NDI)

> **1** This procedure is to establish effective control during performance of non-destructive inspection(NDI) at this repair station. NDI inspection includes Dye Penetrant, Fluorescent Penetrant, Magnetic Particle, Eddy Current, Ultrasonic, X-ray, and Radio Graphic Inspection.

〈단어 연구〉
  Dye Penetrant 염색 침투 탐상, 다이 첵크를 말함
  Fluorescent Penetrant 형광 침투 탐상, 자이글로 테스트(Zygro Test)를 말함
  Magnetic Particle 자분 탐상, 마그너 플럭스(Magnaflux)를 말함
  Eddy Current 와전류 탐상 검사, 와전류에 의한 탐상
  Ultrasonic 초음파 탐상.
  Radio Graphic 라디오 아이소도프를 사용해서 탐상하는 것

〈해 석〉
이 기준은 리페어 스테이션에서 비파괴 검사를 실시할 경우의 효과적인 관리 방법을 설정하는 것이다. 비파괴 검사는 염색 침투 탐상, 형광 침투 탐상, 자분 탐상, 와전류 탐상, 초음파 탐상, X선 및 라디오 그래픽의 각 검사를 포함한다.

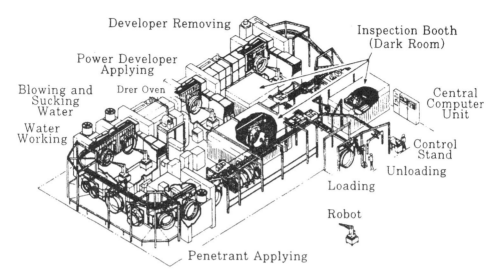

그림 3-8 Automatic Fluorescent Penetrant Inspection System

9) Handling of Parts

> [1] All parts in process through the shop will be properly identified by use of appropriate tags or placed in identified stands, buckets, racks or bins, whichever is appropriate, in order that all parts for each unit will be segregated from other units and protected from damage or contamination.

〈단어 연구〉

in process 진행중
shop 샵, 일반적으로 부품이나 장비품의 정비 공장을 가리킨다
stand 작업대
bucket 버켓
rack 랙, 선반
bin 빈, 칸막이를 한 상자
whichever 어느 쪽이나
segregate 격리하다

134   제3장 Aircraft Maintenance

〈해 석〉
샵에서 정비중인 부품은 모두 적절한 태그를 붙여서 구분해두거나 식별되어 있는 스탠드 버켓, 랙 또는 빈 등 적당한 것에 보관하여 확실히 구별해두어야 한다. 이것은 각 유니트에 사용하는 부품은 다른 유니트와 격리함과 동시에 손상이나 오염을 방지할 필요가 있기 때문이다.

---

② All units repaired or overhauled by this repair station will be preserved in accordance with manufactures' recommendations and standard procedures. The procedures used, dependent upon the units, will be methods which would be appropriate to the parts or units to assure protection until the unit or part is placed into service.

〈단어 연구〉
preserve 보존하다
in accordance with ~에 의해, ~을 바탕으로
dependent upon the units 유니트마다 따로따로
protection 보호

〈해 석〉
이 리페어 스테이션에서 수리하거나 오버홀한 유니트는 모두 제조 회사가 추천하는 방법이나 표준적인 방법으로 보존되어야 한다. 유니트마다 다른 그 보존 방법은 그들 유니트나 부품이 사용되기까지 확실히 보호하는 적절한 방법이어야 한다.

10) Tagging and Identification of Parts

---

① All units for which this repair station is rated, will be properly identified by tags during their progress through the shop. Parts will be tagged as unserviceable until they have completed final inspection when they will be tagged as serviceable.

〈단어 연구〉

3-6. Inspection System of FAA Repair Station

rated 설정되어 있는
unserviceable 사용 불능의

〈해 석〉

이 리페어 스테이션이 확정되어 있는 모든 부품은 샵에서 정비되고 있는 동안은 태그를 달아 확실히 식별해두어야 한다. 또 부품은 최종 검사를 완료하고 사용 가능하다고 하는 태그가 부착될 때까지 사용 불량의 태그를 붙여 두어야 한다.

② The certification for return to service regardless of the type of maintenance performed will be signed by an authorized inspector for the Repair Station. Maintenance release and return to service will not only be limited to maintenance work such as Major Repairs, Major Alterations, but will include Bench Check, Inspection, Minor Repairs, and Preventive Maintenance. Therefore, the Statement on the Serviceable Tag, when signed by an authorized inspector, may not necessarily denote that a major repair, or major alteration has been accomplished. For Major Repair and Major Alterations, the statement will take place of FAA Form 337. If customer requires a FAA Form 337, Repair Station will furnish the customer with FAA Form 337 in lieu of completing the certification statement. In lieu of signature on the certification statement, the authorized inspector will use his assigned stamp.

〈단어 연구〉

regardless 관계 없이
therefore 그러므로
not necessarily 반드시 ~가 아닌
denote 표시하다
take place of ~을 대신하는

〈해 석〉

제3장 Aircraft Maintenance

리턴 투 서비스의 증명은 실시된 정비의 형식에 관계 없이 그 리페어 스테이션을 대표하여 인가된 인스펙터가 서명을 해야 한다. 메인터넌스 릴리즈 및 리턴 투 서비스는 대수리나 대개조와 같은 정비 작업에 한정되는 것 뿐만이 아니고 벤치 첵크, 검사, 소수리 및 예방 정비도 포함된다. 그런 까닭에 사용 가능 태그의 기술은 인가된 검사원이 서명할 경우 반드시 대수리 또는 대개조가 완료되었다는 취지를 표시할 필요는 없다. 대수리 및 대개조에 대해서 그 기술은 FAA Form 337을 대신하는 것이다. 만약 고객이 FAA Form 337을 요구할 경우에는 리페어 스테이션은 증명서를 작성하는 대신 고객에 대해 FAA Form 337을 작성할 필요가 있다. 증명서를 서명하는 대신에 인가된 검사원은 본인에게 지정되어 있는 스탬프를 사용해도 좋다.

11) Inspection Stamps

> ① Purpose : This is to establish procedures for the application, classification, function and issuance of inspection stamps used by authorized inspectors of this repair station.

〈해 석〉
　목적 : 이것은 이 리페어 스테이션의 인가된 검사원이 사용하는 검사인의 적용, 구분, 기능 및 발행에 관한 절차를 수립하기 위함이다.

> ② Scope : These procedures will apply to all parts, components, materials, and equipment upon completion of inspection functions by all authorized inspectors.

〈단어 연구〉
　scope　적용 범위
　upon completion of　~가 완료되어 있는(~가 완료된데 대해서)

〈해 석〉
　적용 범위 : 이들 절차는 인가된 검사원의 검사가 완료된 모든 부품, 장비품, 재료 및 기기에 대해 적용된다.

3-6. Inspection System of FAA Repair Station

3 General Application :
(a) Inspection stamp is a method to indicate the inspection status of an item, not necessarily the acceptance of it.
(b) Inspection stamps are issued and used only by authorized inspection personnel.
(c) A serially numbered stamp will be used to identify the inspector and a activity to which he is assigned.
(d) Assigned stamps must be available for use at all times when the inspector is on duty.

〈해 석〉

적용 일반 :
(a) 검사인은 반드시 품목의 합격을 표시한다고 할 수 없고 그 품목의 검사 단계를 나타내는 하나의 방법이다.
(b) 검사인은 인가된 인스펙터에 대해 발행되며 인가된 인스펙터에 의해서만 사용된다.
(c) 일련 번호가 붙어 있는 검사인은 검사원 및 검사원이 지정되어 있는 행위를 식별하기 위해 사용된다.
(d) 지정된 검사인은 검사원이 근무중일 경우 언제라도 사용할 수 있게 해두어야 한다.

138 제3장 Aircraft Maintenance

3-7. Ground Support Equipment

1) G.S.E. for Maintenance

> ① Towing Tractor : This tractor, conforming to the A.T.A Specifications, is used for towing B747. The engine is diesel and the power train is mechanical. Operator's cabs are provided at both ends of the vehicle. By raising the cab, the operator can obtain good vision in the rear of the vehicle.

〈해 석〉

견인차 : 이 트랙터는 ATA 규격에 합치되었으며 보잉 747을 견인하는데 쓰인다. 엔진은 디젤이고 동력의 전달은 기계적으로 한다. 운전실은 차량의 전후에 설치되어 있다. 운전실을 올리면 운전자는 차량의 후방을 잘 볼 수 있다.

> ② Towing Bar : The towing bar is light and well balanced, and ideal push-pull towing of airplanes in normal ramp use. The coupling level is adjustable with hydraulic hand pump and its towheads are interchangeable.

〈단어 연구〉

　　towing bar 비행기의 견인봉, 토우 바(tow bar)라고도 한다
　　in normal ramp use 통상의 램프에서 사용하는 한에서는
　　coupling level 비행기와 결합하는 레벨
　　towhead 바와 비행기를 결합하는 바

〈해 석〉

토잉 바 : 이 토잉 바는 경량으로 밸런스가 잘 취해져 있으며, 램프상에서 비행기를 밀거나 당기거나 하는데 이상적이다. 결합 부분의 레벨은 수동의 유압 펌프로 조절 가능하며 토우 헤드는 교환 가능하다.

3-7. Ground Support Equipment

3 A.C Ground Power Unit : The generator set is designed to provide a source of 90kVA, 400hertz, 115/200volt, 3-phase, 4 wire, electrical energy for ground support of any aircraft requiring this output. The generator set is a tow tractor mounted, completely self contained, totally enclosed, weatherproof unit. The generator set is provided with all the instruments, controls, and accessories necessary for operation.

〈해 석〉

교류 지상 동력 장비 : 이 발전기는 90kVA, 72kW, 400Hz, 150~200V, 3상 4선의 전력을 이 출력을 필요로 하는 항공기에 공급하기 위해 설계된 것이다. 이 발전기는 견인 트랙터에 탑재되며 완전한 자장식으로 풍우에 노출되어도 괜찮게 완전히 덮여 있다. 이 발전기는 그 운전에 필요한 모든 계기나 제어 장치 및 장비품을 갖추고 있다.

4 Dispenser : This dispenser is used to fuel the aircraft from the hydrant. It is servicing the various types of aircraft. The hydraulically operated elevating platform has hose to protect the hoses from vibration at fueling. Pressure control system is automatic and constructed of 1,000 GPM rate units.

〈단어 연구〉

    hydrant 하이드런트, 급유전(지하 연료 저장조에서부터 연료를 빼쓰는 배출구)
    vibration 진동
    1,000GPM 매분 1,000 갤론(Gallon Per Minute의 약자)

〈해 석〉

디스펜서 : 이 디스펜서는 급유전에서 항공기에 연료를 공급하는데 사용된다. 이것은 여러가지 형식의 항공기에 사용할 수 있다. 유압 작동의 승강대는 급유시에 진동으로부터 호스를 보호하는 호스가 있다. 급유 압력 시스템은 자동으로 매분 1,000gal의 급유 규격이 있는 기기로서 제작되고 있다.

140 제3장 Aircraft Maintenance

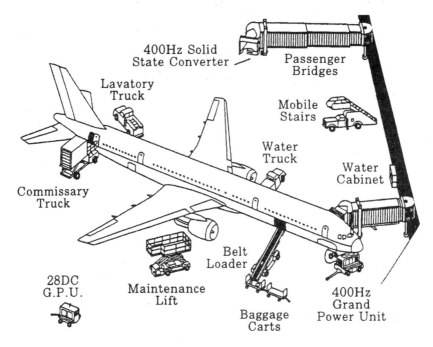

그림 3-9

5 Jacks : The nose and main landing gear contain integral jack pads located on and below each axle, midway between each set of wheels. The type of jack required for nose or main gear jacking varies with purpose and tire condition. With one or both tires normally inflated on any axle, conventional axle jacks of adequate capacity, lift, and body dimensions are satisfactory. Certain flat tire conditions require the use of a cantilever type jack.

〈단어 연구〉

contain 포함하다
integral 내장의, 하나로 된
jack pad 잭 패드, 키를 대는 부분
midway 중간에
flat tire 펑크난 타이어

〈해 석〉

잭(키) : 노스 랜딩기어 및 메인 랜딩기어에는 차축 하면의 차륜 중간 부분에 내장된 잭키 패드가 붙어 있다. 노스 랜딩기어이나 메인 랜딩기어를 들어올리는데 필요한 잭키의 형은 그 목적과 타이어의 상태에 따라 달라진다. 한쪽 또는 양쪽의 타이어가 정상으로 팽창되고 있을 때는 내중량 능력, 리프트, 본체의 치수 등이 적절한 또 종래부터 있던 액슬 잭키로 충분하다. 타이어가 펑크가 났을 때는 캔틸레버식 잭키를 사용할 필요가 있다.

6 Three jack points are provided on the primary structure for jacking the entire airplane. One is located aft of the nose gear wheel well at station 598, 43 inches to the right of the fuselage centerline ; the other two are located one on each wing, outboard of the main landing gear and in line with fuselage station 1465. The jack points provide receptacles for the attachment of removable jack adapters. Tripod jacks are used for jacking the entire plane.

〈해 석〉

비행기 전체를 들어 올리기 위한 3점의 잭키 포인트가 기체의 일차 구조부에 설치되어 있다. 하나는 노우즈 기어 휠 웰 후부, 동체 스테이션 598, 동체 중심선의 우측 43in인 곳에 있고 다른 두개는 각 윙의 메인 랜딩기어의 바깥쪽에 동체 스테이션 1465의 선상에 설치되어 있다. 각 잭키 포인트에는 떼낼 수 있는 잭키 아답터를 끼우는 홈이 있다. 비행기 전체를 들어올리려면 트리포드식 잭키를 사용한다.

2) Stands and Ladders for Maintenance

1 Tailstand for Jumbo Jets : Control of the vertical, horizontal and swing movements of the cantilever work platforms is accomplished from the work platforms. Vertical movements are made by air powered winches with electrical control circuits. The horizontal movements are made by means of air motor driven screw jacks. The swing, of the platforms, is performed by manual drive. The stand is stable in winds to 40mph for all working conditions and

142 제3장 Aircraft Maintenance

against roll-away. It will not overturn in winds to 70mph. For winds above 70 mph it will require tie down cables. Numerous safety features have been built into this stand with operator and maintenance safety in mind, as well as precautions against aircraft damage. A minimum of maintenance is needed for a long service life.

〈단어 연구〉
tail stand 미부 작업대
jumbo jet 대형 젯트기
swing movement 회전 방향의 움직임
winch 윈치
by means of ~에 의해
stable 안정되어 있는
roll-away 굴러 이동하다
overturn 무너지다, 쓰러지다
feature 특징
precaution 예방책

〈해 석〉
대형 젯트기용 미부 작업대 : 캔틸 레버식 작업대를 수직, 수평 및 수평 회전 방향으로 움직이는 조작은 작업대상에서 할 수 있다. 수직 방향으로는 전기 제어 회로가 있는 공기 동력 윈치로 움직여진다. 수평 방향으로는 공기 모터 구동의 스크류 잭키로 움직여진다. 작업대를 수평 회전 방향으로 움직이려면 수동으로 한다. 이 작업대는 모든 작업 상태에서 시간당 40mile의 바람이라도 안정되어 있으며 또 굴러서 이동하는 일은 없다. 풍속이 시간당 70mile까지는 쓰러지는 일이 없다. 시간당 70mile 이상의 바람에서는 케이블로 계류한다. 이 작업대에는 항공기를 파손하지 않게 할 예방책과 동시에 조작을 하는 사람이나 정비상의 안전을 고려에 넣어 여러가지 안전책이 마련되어 있다. 필요 최소한의 정비로 사용 수명을 연장할 수 있다.

2 Maintenance Platform : This platform provides easy positioning and access for work crews to the lower fuselage, electronic compartments, wing leading edge and lower surface, flap and landing gear bay areas, both sides of No.1 and No.3 engines, and all cargo doors. The above working areas are within 16 feet from the ground. Partslikely to come into

> contact with the aircraft are padded to prevent damage to aircraft.

〈해 석〉

정비 작업대 : 이 작업대는 쉽게 위치를 정할 수 있고 작업자가 동체 하면, 전자장비실, 주익 리딩에이지 및 밑면, 플랩, 랜딩 기어 격납실 부근, 1번 및 3번 엔진의 양측, 모든 화물실 도어에 접근할 수 있다. 이와 같은 작업 장소는 지면에서 16ft 이내인 위치에 있다. 기체와 접촉할 것같은 부분에는 기체의 손상을 방지하는 패드가 부착되어 있다.

제4장 AIRCRAFT SYSTEM DESCRIPTION

4-1. Types, Design Features, and Configurations of Transport Aircraft

1) Boeing 737 Series

The 737, shown in Figure 4-1-1, is the smallest member of the Boeing family of jet airliners. a Shortrange airplane, the twin-jet has the same body width and the same accommodations found in other boeing jetliners. The 737-100 and 200 Series, powered by two wing-mounted Partt & Whitney JT8D engines, are capable of flying routes of 100 to more than 2,000miles.

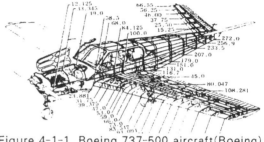

Figure 4-1-1 Boeing 737-500 aircraft(Boeing)

The 737 was ordered into production in early 1965, and the first 737 completed its maiden flight in 1967. The 737-100 is 94 feet long overall and the 737-200 is 100 feet long. The basic dimensions and profile drawings of a 737-200 are illustrated in Figure 4-1-2. The high-lift system of the 737 uses triple-slotted trailing-edge wing flaps, leading-edge slats outboard of the nacelles, and single-type flaps along the wing between nacelles and body. With these high-lift devices, the 737 can operate from runways as short as 4,000 feet.

The 737 was the first Boeing jetliner not requiring a flight engineer. It is operated by a pilot and copilot from a flight deck where simplicity is the keynote.

Because of its role as a short-range jetliner,the 737 is designed so that little ground support equipment and few servicing units are required. Baggage handling and most servicing and maintenance can be accomplished from ground level.

The newest members of the Boeing 737 jetliner "family" are the 737-300, -400, and -500 models. Designed to take advantage of more efficient engines and to carry more passengers than earlier 737's, the 737-400 was ordered into production in 1981.

Design changes resulted in lower direct operating costs and a 20% reduction in fuel

146   제4장 Aircraft System Description

ADVANCED 737-200 SPECIFICATIONS:

Span	93 ft
Length	100 ft
Tail Height	37 ft
Wing Area	980 sq ft
Gross Weight	116,000 lb
Cruising Speed	575 mph
Range	2,000 mi
Service Ceiling	35,000 ft
Power	(2) 14,500-16,000-lb-thrust P&W JT8D-9, -15, or -17 turbofan engines in quiet nacelles
Capacity	130 passengers

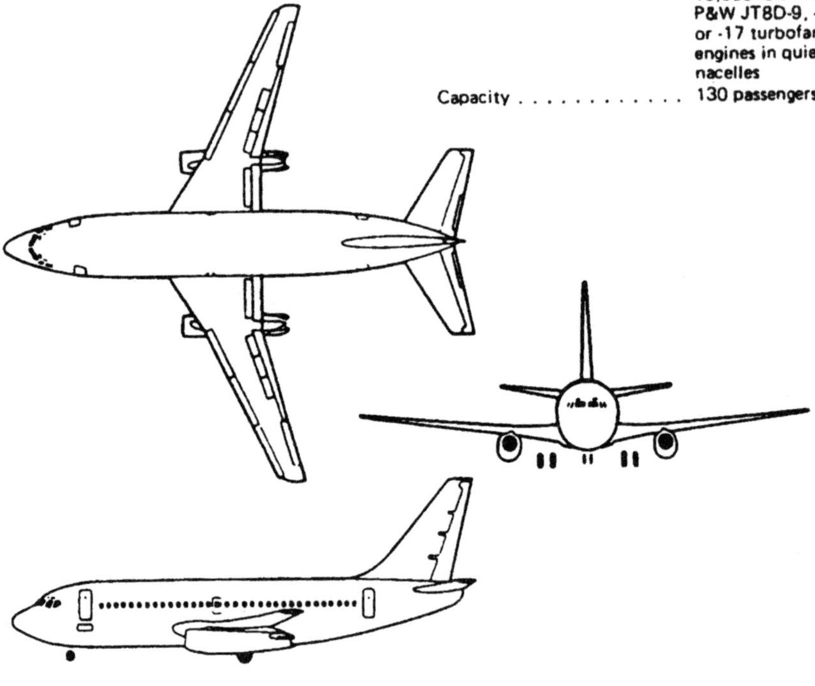

Figure 4-1-2 Boeing 737-200 specifications(Boeing)

burned per seat over the 737-200.

The 737-300 uses two wing-mounted CFM560-3 turbofan engines produced by CFM International, which is jointly owned by General Electric and SNECMA of France. These engines also provide greatly reduced noise levels and have thrust ratings of around 20,000 pounds.

Visible differences from the standard 737-200 are the 104 inch(2.6m) fuselage extension, giving the airplane an overall 9 foot 7 inch(33.4m) greater length, and the larger engines mounted forward of the wing on struts(737-100 and -200 engines are tucked up

4-1. Types, Design Features, and Configurations of Transport Aircraft 147

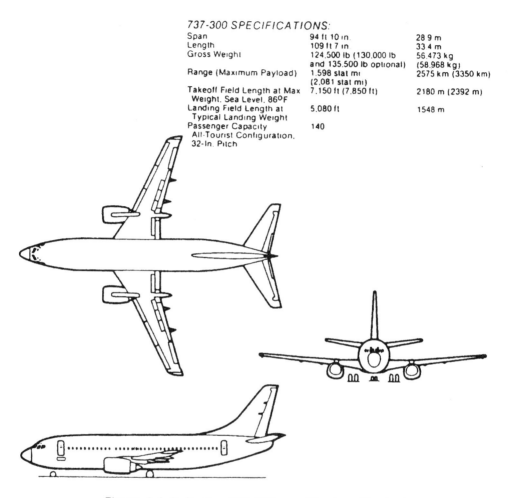

Figure 4-1-3 Boeing 737-300 specifications (Boeing)

directly under the wing). The new 737-300 also has 6 feet of span added to the horizontal tail and small wing-tip extensions. Specifications and a drawing of the 737-300 is shown in Figure 4-1-3.

The flight deck reflects digital technology instrumentation. This equipment provides 737 operators with substantial commonality benefits in equipment. The 737-300, although having 80% commonality with the present 737-200, is considered a new generation airliner because of the use of composites inside and out as well as the new engines and avionics.

The 737-400 is 120 inches longer than the 737-300 and has a maximum takeoff weight of 142,500 pounds, with increases in gross weight expected in the future. The engines are CFM56-3B-2's or -C-1's rated at 22,000 pounds SLST(sea level static thrust). The aircraft will carry between 135 and 172 passengers, depending upon the seating arrangement. The 737-400 is the largest of the 737 family.

The 737-500 began service in 1990 for lower density short-to-medium range routes. The 737-500 incorporates the proven technology features of the 737-500 is offered in gross weights from 115,500 pounds up to 133,500 pounds, giving it the capability to carry 108 mixed class passengers 2,500 nautical miles. The engines are CFM56-3-B1's rerated to 18,500 to 20,000 pounds SLST. Other than the length differences, the 737-300, -400, -500, aircraft are very similar with regard to basic shape and engine location.

2) Boeing 747 Series

Design of the Boeing 747, first of the giant jetliners, began in the early 1960's, when market research indicated the need for a much larger capacity transport to cope with the growing passenger and cargo traffic. In 1968 and the first scheduled 747 service began in 1970. A747-400 aircraft can be seen in Figure 4-1-4.

Figure 4-1-4 Boeing 747-400 aircraft(Boeing)

The 747 has been produced in several basic models, the 100, -200, -300, 747SR, 747SP and -400 Series with the -400 Series being the latest version. The different model series and the first flight dates are shown in Figure 4-1-5.

The 747 is offered in three basic configurations:all passenger, mixed passenger/cargo, and all freighter. In convertible or freighter versions, the 747's volume and upward hinged nose allow straight in loading of long and bulky articles. An optional main deck cargo side door is available on all three models.

An unusual feature of the 747 is the unique upper deck, immediately behind the cockpit. This area can be used as a luxurious first class lounge, or for standard passenger seating.

The wing flap system and a 16-wheel, four-maintruck landing gear allow the 747 to operate from runwyas normally used by 707's. As more powerful engines became available, new range and load capabilities became possible. The initial 747-100 had a 710,000-pound takeoff weight, which has been raised to 833,000 pounds for later 747's.

4-1. Types, Design Features, and Configurations of Transport Aircraft   149

747 Model Series	First Flight Date
747-400M	1989
747-400	1988
747-300M	1982
747-100B	1979
747-SP	1975
747-200B/C	1974
747-SR	1973
747-300	1973
747-200L	1973
747-200F	1971
747-200	1970
747-100	1969

Figure 4-1-5 First flight dates and Boeing 747 model series (Boeing Airliner)

In 1972, the first 747F Freighter went into service, carrying loads of more than 200,000 pounds in regular service, with the first of three 747C Convertibles beginning service in 1973. The 747SR(Short Range) began service with Japan Air Lines, seathing 498 passengers. Later, seating capacity of newly delivered 747SR's increased to 550. The 747 family has grown to included several different models. Specially equipped Advanced Airborne Command Post 747's, designated E-4, replaced U.S. Air Force EC135's. Boeing also modified several 747-100's as combined tanker transports, special freighters, and the vehicle to transport the Space Shuttle.

A. 747-200B Specifications
Span ---------------------- 195ft 8 in
Length --------------------- 231ft 10 in
Tail Height --------------- 63ft 6in
Wing Area ---------------- 5,500sq ft
Gross Weight ------------ 775,000~833,000lb
Crusing Speed ----------- 600 mph plus
Range --------------------- 6,000nm
Design Ceiling ----------- 45,000ft
Power --------------------- (4) P&W JT9D(46,950 to 54,750 lb thrust) ;

150 제4장 Aircraft System Description

        G.E. CF6-50E2(52,000lb thrust)
        Rolls-Royce RB211(50,100~53,110 lb thrust)
Typical Camacity ---------- 452 passengers
Fuel Volume -------------- 52,409~53,985 U.S.gal

B. 747 Design Features

The 747-400 is a wide body airplane with four wing-mounted engines of the 747-400 and is designed for long range operation at high payloads. The maximum range is approximately 7,300 nautical miles(with horizontal stabilizer fuel tnak). The aircraft is derivative of the 747 family(-100, -200, -300) and includes major new features of:

- Two-Crew flight deck. The earlier models of the -747 required a flight engineer to control aircraft systems.
- Crew rest area
- Advanced avionics and electronics
- High performance engines
- Advanced APU
- Wing tip extensions and winglets

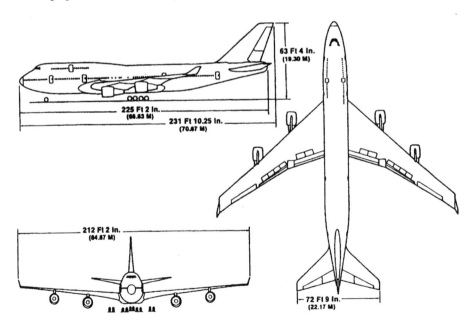

Figure 4-1-6 Dimensions of the Boeing 747-400(Boeing)

4-1. Types, Design Features, and Configurations of Transport Aircraft 151

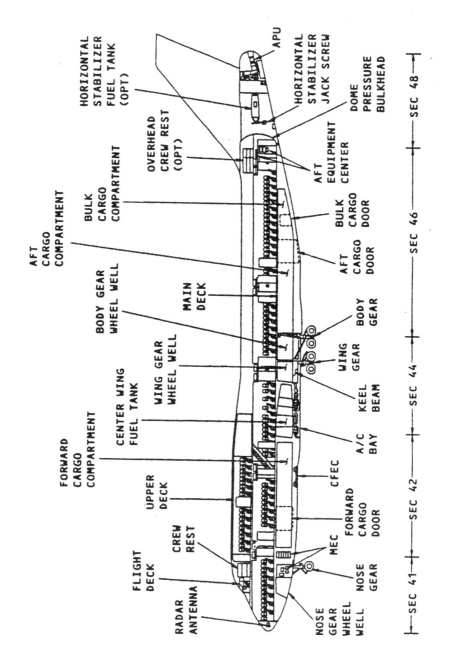

Figure 4-1-7 General arrangement of a Boeing 747-400(Boeing)

152 제4장 Aircraft System Description

- Advanced aluminum alloy wings
- Optional horizontal stabilizer fuel tank
- Increased gross weight
- Interior configuration flexibility
- Carbon brakes

The principle dimensions of the 747-400 airplane are shown in Figure 4-1-6 with a frontal and profile view. The maximum width of the airplane is the wing span of 212 feet nominal, maximum is 213 feet, which includes the winglets. The wing and horizontal stabilizer are set at a 7 degree dihedral. The wing dihedral affects engine mounting because the struts are attached perpendicular to the wing. The maximum height of the airplane, 63 feet 4 inches, is from the ground to the vertical stabilizer tip. The vertical stabilizer is swept back 45 degrees. The maximum body height of the airplane, 32 feet 2 inches, is from the ground to the upper deck skin. The maximum length of the airplane, 231 feet 10 inches, is from the radome to the vertical stabilizer tip. The general arrangement of the 747-400 is shown in figure 1-7.

C. 747 Powerplants

Engines from General Electric, Pratt & Whitney, and Rolls-Royce are all available for the 747 and each manufacturer has met the challenge of developing powerplants with ever increasing thrust ratings. Engine certification evolution is shown in Figure 4-1-8.

Most of the engines certificated for the 747 series of aircraft are in the area of 50,000 to 60,000 pounds of thrust. A cutaway picture of a typical high bypass engine used on the 747 is illustrated in Figure 4-1-9.

Manufacturer	Engine	Year Certified
Rolls-Royce	RB211-524G	1989
General Electric	CF6-80C2B1F	1989
Pratt & Whitney	PW4056	1988
General Electric	CF6-80C2B1	1987
Pratt & Whitney	JT9D-7R462	1983
Rolls-Royce	RB211-524D4	1981
Pratt & Whitney	JT9D-7Q	1979
Rolls-Royce	RB211-524B2	1979
General Electric	CF6-50E2	1979
General Electric	CF6-45A	1978
Rolls-Royce	RB211-524	1977
Pratt & Whitney	JT9D-70A	1976
Pratt & Whitney	JT9D-7F	1975
General Electric	CF6-50	1974
Pratt & Whitney	JT9D-7	1971

Figure 4-1-8 Certification avolution for Boeing 747 engines(Boeing Airliner)

4-1. Types, Design Features, and Configurations of Transport Aircraft   153

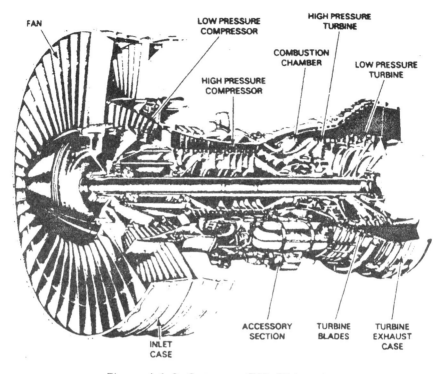

Figure 4-1-9 Gutaway JT9D-7R4 engine

3) 767 Twin Jet

The Boeing 767 shown in figure 1-10, is a new generation commercial airplane whose design makes use of the latest in technology to provide maximum efficiency. It provides modern twin-aisle wide-body passenger cabin comfort.

Figure 4-1-10 A Boeing 767 Aircraft(Boeing)

Production fo the new twin-jet began in 1978. The 767's design has been refined to give maximum performance per unit of fuel burned. Design achievements also include operational flexibility, low noise levels, advanced airplane systems, and growth potential. New structural material, including advanced composites, are also employed.

154 제4장 Aircraft System Description

The 767 cabin, measuring more than 4 feet wider than the standard fuselage in Boeing jetliners now in service, will seat 211 passengers in a typical mixed-class configuration (six-abreast in first class, seven-abreast in tourist class), or as many as 290 in the all-tourist configuration.

The first 767 was rolled from the factory in 1981 with Pratt & Whitney engines. A second 767 type, powered by General Electric CF6 engines, also received FAA certification. Another variation of the basic model, the 767-200ER(Extended Range)version was certified with a higher gross weight. A longer body version, the 767-300, was also certified in 1983.

A. 767-200 Specifications

Span ------------------ 156ft 1in
Length ---------------- 159ft 2in
Tail Height ----------- 52ft
Wing Area ------------- 3,050sq ft
Body Width ------------ 16.5ft
Passengers ------------ 211(mixed class);up to 290 in charter configuration
Cruising Speed -------- Mach .80
Lower Deck Volume ----- 3,102cu ft
Gross Weight ---------- 300,000 to 345,000lb
Range ----------------- 3,200 to 5,600 stat mi
Engine ---------------- Two Pratt & Whitney
                        JT9-7R4 or General Electric
                        CF6-80A, at airline option
                        (48,000-lb thrust)
First Delivery -------- August 19, 1982

B. 757-767 Design Features

The two crewmember airliner flight deck, shown in figure 1-11, is standard on the new-generation Bowing 757 and 767 twin-jets. The flight deck features digital electronics, including an Engine Indicating and Crew Alerting System(EICAS) that centralizes all engine displays and provides automatic monitoring of engine operation. Taking advantage of equipment commonality, the 757 and 767 flight decks feature the same elements. forward windshields are identical, yielding downward visibility during landing approach.

4-1. Types, Design Features, and Configurations of Transport Aircraft   155

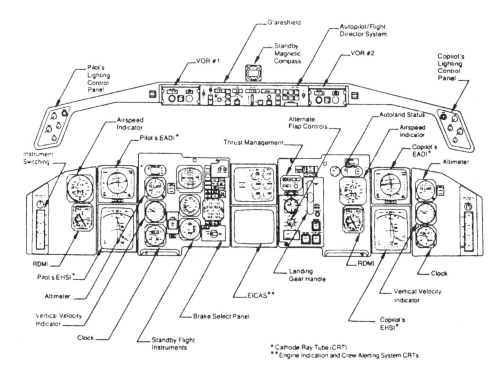

Figure 4-1-11 Flight deck instruments Boeing 767 (Boeing)

The flight decks employ digital electronic instruments, including color cathod-ray tube displays.

An all-digital electronics automatic flight management system not only reduces crew workload, but also contributes to lower fuel consumption. Use of digital electronics also makes possible upgrading in navigation capability as air traffic control systems change. The flight deck includes a low-profile control column designed to permit full view of the instrument panel. Design of the nose section and location of air-conditioning outlets means lower cockpit noise levels.

The cargo system consist of a forward, aft and bulk cargo compartments. A complete cargo loading system is available for lower deck cargo as shown in Figure 4-1-12.

The latest Boeing airplane to be proposed is the 767X or the 777, which at the time of this printing is still in the planning and engineering stages.

156 제4장 Aircraft System Description

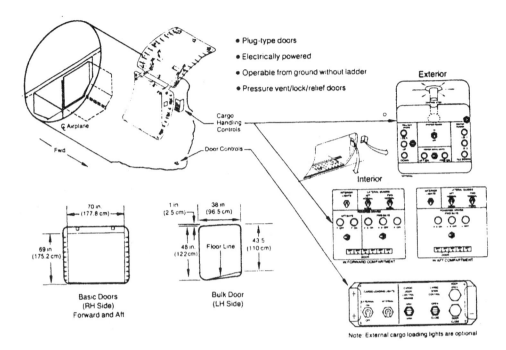

Figure 4-1-12 Cargo loading system Boeing 767(Boeing)

C. 767 Powerplants

The 767 can be equipped with a Pratt & Whitney JT9D-7R4 or a General Electric CF6-80 series engine, with other engine, with other engine types also being produced. Takeoff thrust of these series of engines ranges from 48,000 to 61,500 pounds, allowing airline operatiors to select the appropriate thrust level for aircraft and route requirements.

Thrust of the new PW4000 engine series ranges from 50,000 to more than 60,000 pounds, with growth capability to 70,000 pounds, making the engine ideal for all current and new versions of the Boeing 747 and 767.

4) McDonnell Douglas MD-80 Series

The McDonnell Douglas MD-80, a quiet and fuel-efficient twin-jet, uses advanced Pratt & Whitney JT8D-200 engines. The liberal use of sound-suppressing materials allows the MD-90 to meet stringent noise regulations.

4-1. Types, Design Features, and Configurations of Transport Aircraft　157

Figure 4-1-13　McDonnell Douglas MD-80 aircraft(McDonnell Douglas)

Five MD-80 models, the MD-81, MD-82, MD-83, MD-87, and MD-88 are produced at the Douglas Aircraft Company's Long Beach, California facility. The MD-81, -82, -83, and -88 are 147 feet 10 inches(45.05 meters) long and accommodate a maximum of 172 passengers. The MD-87 is 130.4 feet(39.7m) in length, with a maximum passenger capacity of 139. Wingspan for all models is 107 feet 10 inches(32.86m).

Technology advancements in the MD-80 include a digital flight guidance system with a nonstop range of 1,750 to 3,260 statute miles(2,815 to 5,246 kilometers), depending on the model. the MD-80's maximum takeoff weight ranges from 140,000 pounds(63,503kg) to 160,000 pounds(72,576 kg).

The area of the MD-80 wing is 1,209 square feet(112.3sq m), approximately 21 percent greater than the 1,000.7 square feet(92.9sq m)of the DC-9. With the larger wing, MD-80 fuel capacity totals 5,779 gallons(21,873.5 L), an increase of 2,100 gallons(7,949 L) over the standard capacity of the DC-9 Series 30. In the MD-83, fuel capacity is 7,000 gallons(25,500 L).

Height of the MD-80's distinctive T-tail is 29 feet 8 inches(9.02 m). The horizontal stabilizer has a span of 40 feet 2 inches(12.2 m). The MD-87 has a higher vertical tail, 30.5 feet(9.29 m).

A typical cabin arrangement in the MD-80 is 142 seats, with a first class section of 12 seats and a coach area of 130. Other interior arrangements include 155 all economy-class seats and up to 172 seats in MD-80's intended for short route service. On the MD-87's, seating capacity ranges from 100 to 130. Seats are in a four-abreast pattern in first class and five-abreast in coach and economy sections.

The three belowdeck cargo compartments on MD-81, MD-82, and MD-88 models have a total volume of 1253 cubic feet(35.48 cu m). Cargo is loaded into the compartments through thre doors(two forward, one aft), each 50 × 53 inches(127 × 134.6 cu m).

제4장 Aircraft System Description

A. MD-80 Powerplants

Power for the MD-80 is generated by two Pratt & Whitney Aircraft JT8D-200 series turbofans, mounted one on each side of the aft fuselage. The MD-81 was certified with the JT8D-217A and -217C engines for operations from high altitude, high-temperature airports are available. The JT8D-217C or the JT8D-219 engines for performance from short airfields over extended ranges. The JT8D-219 is rated at 21,000 pounds(79,493 kg) plus 700 pounds(317.5 kg) in reserve. both the -217C(MD-88) and the -219 engines offer fuel burn reductions of about two percent through use of advanced technology hardware. The larger fan increases the bypass ratio from about 1.00 for earlier JT8D engines to about 1.7, resulting in lower specific fuel consumption and noise emission.

B. MD-80 Design Features

The two-person flight crew allows for advanced technology systems which are designed into the MD-80. Electronic flight instrumentation(EFIS), wing shear detection capability, advanced flight management systems for improved navigation, and more fuel effcient flight profiles are offered on the MD-80 Series. In addition, structural life has been extended to more than 50,000 landings with added protection against corrosion. The MD-80's integrated digital cockpit improves aircraft performance and reduces fuel consumption. for example, the auto-throttles can be engaged prior to takeoff and the autopilot, shortly after takeoff. Once these systems are engaged, the pilot is not required to touch the manual flight controls and throttles unitl rolling out on the ground after landing.

The MD-80 is equipped with two identical Sperry digital computers to direct seven digital flight control subsystems: the automatic pilot, speed control system, automatic throttles, thrust rating system, altitude alert, flight director control function and the automatic reserve thrust system. Either computer has the capacity to independently manage the subsystems.

The MD-80 is equipped with an automatic reserve thrust system that functions during takeoff in case one of the two engines should fail. Detecting the loss of thrust, the flight guidance computer automatically signals the other engine to increase its thrust to compensate for the decrease in takeoff power.

C. MD-83

The MD-83 carries 155 passengers and their baggage 2,900 statute miles(4,670 km) with a maximum seating capacity of 172 passengers. Maximum takeoff gross weight is

160,000 pounds(72,576 kg). The extra range of the MD-83 comes from the takeoff weight increase and installation of two fuel tanks in the belly of the aircraft, increasing the fuel capacity by 1,160 gallons(4,290 L). This gives the MD-83 a total fuel capacity of 7,000 gallons(26,495 L).

D. MD-88

Development of the MD-88, newest of the popular McDonnell Douglas MD-80 series of twin-jets, began in January 1986. Featuring advanced interior design and cockpit systems, the MD-88's first flight and certification was in 1987. Standard powerplants for the MD-88 are two Pratt & Whitney JT8D-217C Engines, each of which develops 20,000 pounds(9,072 kg) of takeoff thrust. with a full load of passengers and baggage, range of the MD-88 is more than 2,360 statute miles(3,800km).

5) Douglas DC-10 Series

The excellent performance of the DC-10 tri-jet, pictured in figure 1-14, is demonstrated by its operational statistics. Since its introduction into scheduled airline service in 1971, the DC-10 has flown more than 7.1 bilion miles(10.94 billion km) in revenus service.

Five versions of the DC-10 are in airline service. The Series 10, equipped with General Electric CF6-6 turbofan engines, has a maximum range of 4,400 miles(7,079 km). The Series 15, similar to the Series 10, offers additional thrust for takeoffs from hot climate, high-altitude airports. Two intercontinental models are the Series 30(Figure 4-1-15), powered by General Electric CF-6-50 engines. and the series 40, with Pratt & Whitney aircraft JT9D engines.

The fifth model of the tri-jet is the DC-10CF(convertible freighter) which can be arranged to carry all-cargo, all-passengers or a combination of the two. All models of the DC-10 will carry up to 380 passengers, but most airlines arrange the interior for 250 to 275 seats in the first class and coach arrangement. The DC-10 has an almost 19-foot-wide(5.7 m) cabin.

Figure 4-1-14 DC-10 aircraft (McDonnell Douglas)

160 제4장 Aircraft System Description

A. Series 10

The DC-10 Series 10 model, with transcontinental U.S. non-stop range, was the first of the McDonnell Douglas wide-cabin tri-jet series. Designed for service on routes of 300 to 4000 statute miles(480 to 6,436 km), the Series 10 is powered by General Electric CF6-6 engines, each rated at 40,000 pounds(17,144 kg) takeoff thrust.

Figure 4-1-15 McDonnell Douglas DC-10-30 aircraft(McDonnell Douglas)

B. Series 40

The second DC-10 model launched was the intercontinental range Series 40, built in two versions. One, with a range of 5,350 miles(8,608 km) is powered by Pratt & Whitney JT9D-20 turbofan engines. The other, equipped with the more range of approximately 5,800 miles(9,322 km).

C. Series 30

The Series 30, an intercontinental version with a range of approximately 5,900 miles (9,493 km), was the third DC-10 model to be committed to production. It is equipped with advanced General Electric CF-6-50 fanjets. The Series 30 began its flight test program on June 21,1972.

D. DC-10CF and DC-10F

The fourth basic DC-10 version is the DC-10CF(convertible freighter). Available in the basic Series 10, Series 30, or Series 40, the DC-10CF uses either the P&W or G.E. engines, and is capable of overnight conversion from a passenger configuration to an all-cargo arrangement and vice versa. All versions of the DC-10CF convertible transport have a total available cargo space of more than 16,000 cubic feet(452.8 cu m), or as much as four 40-foot-class(12.19 m) railroad freight cars.

The main cabin will accommodate 30 standard 88 × 108 inch(223.5 × 274.3 cm) pallets or 22 standard 88 × 125 inch(223.5 × 317.5 cm) pallets. A cargo door, 8 $\frac{1}{2}$ feet × 11 $\frac{1}{2}$ feet(2.6m × 3.5 m), swings upward from the side of the forward fuselage.

E. Series 30F

The DC-10 Series 30F, which is a pure freighter version, will carry palletized payloads of up to 175,000 pounds(79,380 kg) more than 3,800 miles(6,115 km). It can be operated up to 460 miles(741 km) farther with the addition of auxiliary fuel tanks in the aft end of the lower cargo compartment.

F. Series 15

The Series 15 model of the DC-10 combines the basic smaller airframe of the Series 10 with a version of the more powerful engines used on the longer range Series 30's. The combination gives the Series 15 outstanding performance with full loads from high altitude airports in warm climates.

It is powered by three General Electric CF6-50C2-F engines, each with a takeoff thrust of 46,500 pounds(21,092 kg). Range of the Series 15, with full payloads, is about 3,600 miles(5,792 km).

G. DC-10 Design Features

Overall length of all DC-10 versions is approximately 182 feet(55.47m). Wingspan is 155 feet 4 inches(47.37 m) on the Series 10 and Series 15 aircraft and 165 feet 4 inches(50.42 m) on the Series 30 and 40. Wings are swept at an angle of 35 degrees. Gross takeoff weight ranges from 440,000 pounds(195,048 kg) for the Series 10, to 580,000 pounds(251,748 kg) for the Series 30 and Series 40.

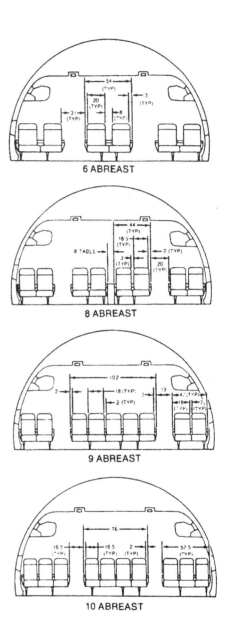

Figure 4-1-16 DC-10 main cabin cross section(McDonnell Douglas)

162   제4장 Aircraft System Description

6) McDonnell Douglas MD-11

The MD-11, shown in Figure 4-1-17, is an advanced wide-cabin, tri-jet aircraft which is an advanced version of the DC-10. The MD-11 is produced by the Douglas Aircraft company. The standard MD-11 is available in three models-passenger, all-freighter and "combi," where passengers and freight are carried on the main deck.

Figure 4-1-17 MD-11 aircraft
(McDonnell Douglas)

Advances in aerodynamics, propulsion, aircraft systems, cockpit avionics and interior design have greatly improved performance and operating economy of all MD-11 models. A wide variety of interior arrangements is available with seating capacities on the standard airplane ranging from 250 in a three-class arrangement to more than 400 in an all-economy configuration.

Below deck, the MD-11 provides combined containerized or palletized cargo capability. Maximum range carrying 323 passengers is up to 8,039 statute miles(12,938 km) nonstop, an improvement of 27 percent compared to the DC-10 Series 30 configured in a comparable two class arrangement.

A. MD-11 Powerplants

Three General Electric CF6-80C2, Pratt & Whitney 4460, or Rolls-Royce RB211-524L engines power the standard MD-11. These engines provide maximum efficiency in their thrust class with substantial fuel burn reductions. The engines are located in much the same manner as on the DC-10 aircraft.

B. MD-11 Design Features

The addition of winglets at the wing tips, redesigned wing airfoils with more camber near the trailing edge, and a smaller horizontal tail with an extended tail cone, help reduce drag. Together with the advanced engines. the aerodynamic improvements lower fuel burn by about 20 percent per trip, compared to the DC-10 Series.

The MD-11's all new flight deck, with advanced systems, makes it possible for operation with a two-pilot crew, compared to the three crewmembers required for earlier

4-1. Types, Design Features, and Configurations of Transport Aircraft

tri-jets(DC-10). The flight deck features six cathode raytube displays, digital instrumentation, wind shear detection, guidance devices, and a dual flight management system that helps conserve fuel. The flight deck also has a dual digital automatic flight control system(autopilot).

The navigation and display systems have double backups for maximum safety. The MD-11 flight deck operates on the "dark cockpit" philosophy-99 percent of the time all lights for switches, warnings, and other devices are off. A light goes on only as an alert, assuring prompt pilot attention.

Status of the aircraft and all its systems is provided to the crew without need for looking up at overhead panels. All alert information is displayed on the engine CRT. The normal, abnormal and emergency checklist functions are performed automatically rather than simply annunciated to the crew, as on previous aircraft.

C. Combi, Freighter Interiors

The longer fuselage of the standard MD-11 has enabled Mcdonnell Douglas to design an efficient "combi" Aircraft(See Figure 4-1-18) that permits a variety of cargo/passenger configurations on the main deck. A typical combi interior, seen in Figure 4-1-19, provides for three-class seating for 176 passengers and six pallets below-deck. A large, main deck rear cargo door for the combi is another MD-11 feature. The MD-11 also can be configured as an all-freighter. Weight limited payload is 205,700 lb(93,304 kg) and non-stop range is 4,076 miles(6,560 km) at maximum payload and landing weight.

Figure 4-1-18 MD-11 combi aircraft (McDonnel Douglas)

D. Automatic System Controllers(ASC)

Automatic system controllers(ASC) are used to control the four basic systems-hydraulic, electrical, environmental, and fuel. The ASC's have replaced the flight engineer. The ASC's are contained in the overhead panel, viewable and reachable by either of the two pilots. The ASC contains four segregated panels, each of which is controlled by two redundant computers and is independent of the other systems. There is manual backup in case of dual computer failures.

164 제4장 Aircraft System Description

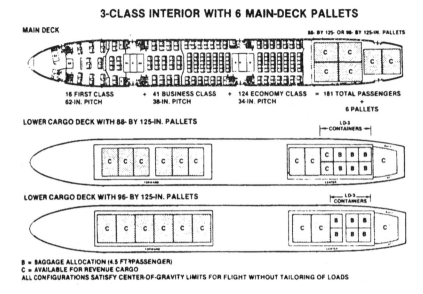

Figure 4-1-19 A typical MD-11 combi configuration(McDonnell Douglas)

The overhead panel is designed so that a logical sequence of aircraft items-hydraulic, electrical, environmental and fuel, are arranged with each engine's systems grouped vertically under its fire shoutoff handle. Thus, it is simple to check or operate any system function.

Automated checklists for all four major systems are completed simultaneously, rather than in series.

E. Flight Control Computer(FCC)

The flight control computer(FCC), formerly called automatic pilot and flight director, now includes automatic throttles and the longitudinal stability augmentatton system(LSAS). LSAS enhances the pitch stability of the aircraft. An automatic landing system that meets FAA Category Ⅲ B all-weather requirements, including rollout, further reduces crew workload and enhances the operational reliability of the MD-11.

The digital FCC interfaces with the full authority digital electronic control(FADEC) engine system, making engine operation precise and smooth. Wing shear detection and guidance data to assist flight crews in escaping shears also are provided by the flight control computer.

4-1. Types, Design Features, and Configurations of Transport Aircraft   165

F. Dual Flight Management System(FMS)

The dual flight management system(FMS) is derived from earlier first generation FMS designs. It includes a larger memory, faster response time and makes management of the flight considerably simpler for the pilot. Designed for optimum efficiency, the FMS guides the aircraft in vertical and lateral flight.

One major workload reduction is the automation of navigational radio management by FMS. The pilot no longer needs to set a variety of headings and radio frequencies during the flight. The FCS and FMS have been functionally integrated so that pilots essentially control speed, heading and pitch on one set of controls that provide maximum capabilities for aircraft operation with minimum workload.

Two multi-function display units for the FMS are on the pilot's console. They also can be used for standby navigational displays. A third display on the console unit provides readouts for such functions as ground-to-air and air-to-ground communication and a display of gates available at the destination airport;it also interfaces with the aircraft's centralized fault-display system, thus providing a single point for maintenance crews to check systems following a flight.

7) Airbus A320

The A320 is a short/medium range twin-engine subsonic commercial transport aircraft introduced as the first single aisle aircraft to the Airbus family. The seating capacity varies between about 120 and 179 passengers.

The design combines the high technology available today with the wide experience gained by Airbus Industries from the A300 and A310.

The Airbus A310-210 use two CFM56-5A1 engines, while the A320-230 uses two V2500-A1 engines. Both of these engines are equipped with a Full Aughority Digital Engine Control(FADEC) system. The engines are wing mounted and are attached by an engine pylon consisting of a main frame which can be converted to accept either of the engine types available for the A320.

The A320-200 Series general dimensions and engine locations are shown in Figure 4-1-20. The aircraft features a two crew flight deck, along with CRT displays and electrically signaled flight controls. The centralized maintenance system and the use of composites in the structure are also advanced features of the A320.

166 제4장 Aircraft System Description

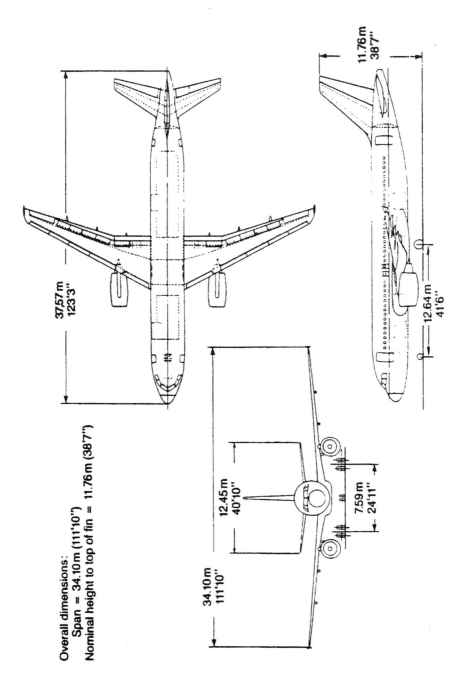

Figure 4-1-20 The general dimensions of the Airbus A320(Airbus)

4-1. Types, Design Features, and Configurations of Transport Aircraft 167

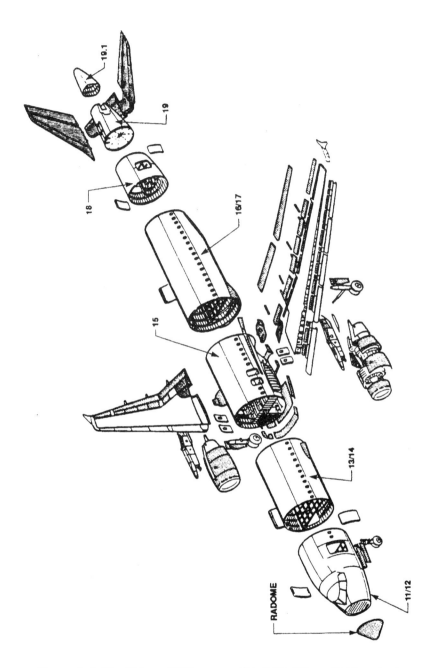

Figure 4-1-21 Airbus A320 composite material application and major structural sections (Airbus)

A. Design Features

The structure of the A320 is generally of conventional design and construction, similar to that employed on the earlier Airbus aircraft. Conventional materials are employed for much of the airframe structure, with more extensive use of improved aluminum alloys.

Composite material has been employed in many areas of the A320 structure, as can be seen in Figure 4-1-21. The fuselage employs a conventional type of skin, stringer and frame construction, except in the nose section where frames are used without stringers. Skin thickness variations are produced by chemical or mechanical machining and the stringers are attached by rivets. All areas of the fuselage are pressurized except for the radome, the rear fuselage section, the nose landing gear bay, and the lower segment of the center section. Provisions have been made for rapid decompression in the event of damage to the pressurized fuselage.

The wing, which is of conventional structure and material, consists of three main components: a center wing box integral to the center fuselage, to which are attached the left and right cantilevered outer wing sections, thus providing a continuous wing structure. Each outer wing section has five slats, two flaps, five spoilers and an aileron. All moving surfaces, except slats, are made of composite materials.

The flight compartment arrangement includes stations for the Captain, First Officer, and a third occupant facing forward. Controls and indications for systems are provided on the main instrument panels, center pedestal and overhead panel.

The passenger cabin, shown in figure 1-22, has a maximum width of 145.5 in(3.70 m) and a height of 84in(3.95 m). However, the cabin height is reduced in the forward and aft entrance areas for technical purposes. The A320 cabin interior has been designed according to the latest industrial design concepts to create an air of spaciousness.

The four seat tracks running the length of the cabin allow 4 or 6 abreast seating. Standard and optional lavatory and galley positions are provided at each end of the cabin on either side of the entrance areas. The seat tracks are spaced apart so as to give approoximately equal space under each seat for carry-on baggage.

Galleys and lavatories are attached to the aircraft structure and are manufactured as pre-assembled cells with an integrated floor panel. This design helps protect against corrosion caused by leakage in these areas.

The main components of the semi-automatic cargo loading system, which are shown in Figure 4-1-23, are available as a standard option. The system is based on a new container dervied from the widely used LD3 container by reducing its height from 64 inches to 46

4-1. Types, Design Features, and Configurations of Transport Aircraft   169

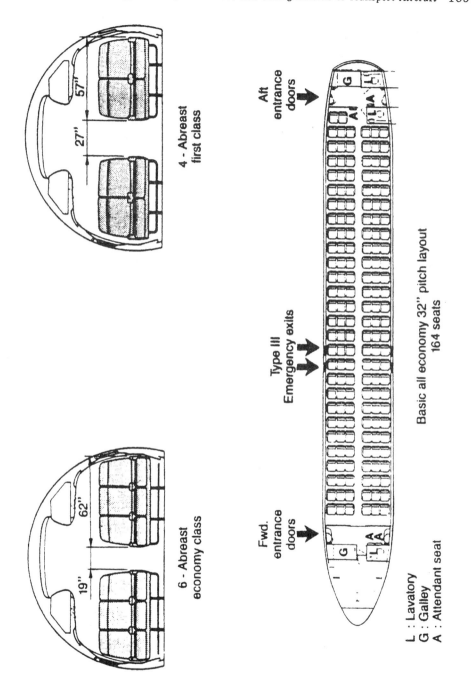

Figure 4-1-22 The A320 cabin furnishings(Airbus)

170 제4장 Aircraft System Description

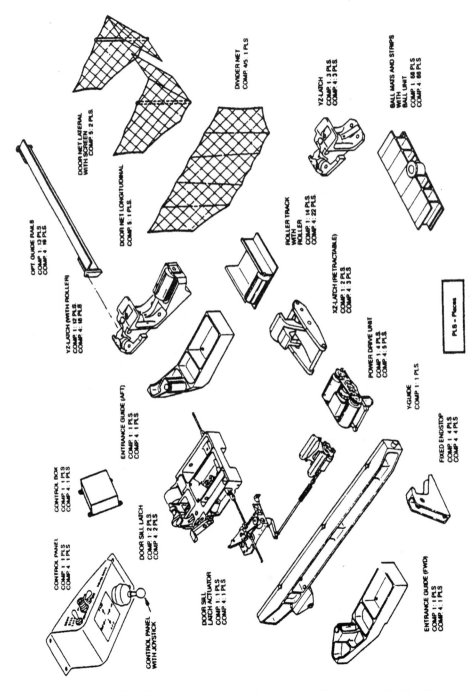

Figure 4-1-23 Main components of cargo loading system(Airbus)

4-1. Types, Design Features, and Configurations of Transport Aircraft   171

inches. The base plate is identical to that of the LD3 container and hence existing container loaders, ground support equipment and warehouse facilities are compatible with the A320 container. In addition, this container can be used in wide body aircraft, providing their system has adequate restraints. The containers are supported on a single roller track along the cargo compartment center line, and on their outer edges, by rollers built into the restraint fittings. Power drives are incorporated into the center roller track with the doorway areas controlled by ball mat and transverse drive systems. The Control Panel, located at the Cargo Loading Door(LD), uses a joystick to activate the Power Drive Units(PDU).

B. A320 Powerplants

The CFM56-5-A1 engine is a derivative of the CFM56 family certified at 25,000 pounds take off thrust. The Full Aughority Digital Engine Control(FADEC) allows "custom tailored" thrust ratings for the optimization of aircraft performance. Fuel consumption versus flight profile and aircraft weight allows optimum functional integration with the A320 Fly-By-Wire Control System. The CFM56-5 engine's modular construction, is illustrated in Figure 4-1-24. The modules have provided for the use of a reduced number of parts, improved sub-assembly designs and better repairability. To achieve a high degree of modular interchangeability, all engine mating points are dimensionally controlled and modular balancing is used.

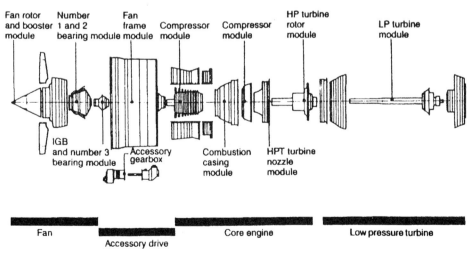

Figure 4-1-24 CFM 56-5 modular engine design(Airbus)

172 제4장 Aircraft System Description

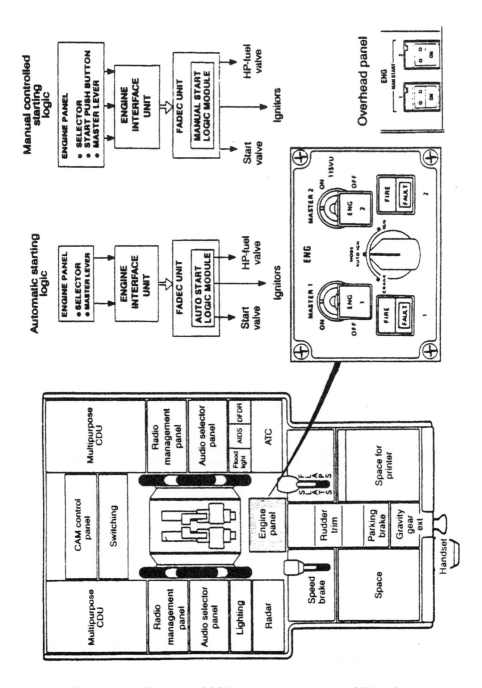

Figure 4-1-25 Airbus A320 main engine starting(Airbus)

4-1. Types, Design Features, and Configurations of Transport Aircraft

The main engine panel is located on the center pedestal in line with the engine throttle levers, as shown in Figure 4-1-25. A320 main engine starting is normally performed using the FADEC automatic start sequencing logic. The rotating selector is positioned in the "IGNITION" position, then the master lever of the corresoponding engine is moved to "ON". The FADEC then automatically, through its start command logic, sequences the starter valve operation, ignition and high pressure(HP) fuel valve. The FADEC then monitors N1, N2, and EGT and ensures the appropriate limit protection. A manual override function allows start sequence to be controlled by the pilot via FADEC.

174 제4장 Aircraft System Description

4-2. Auxiliary Power Units, Pneumatic, and Environmental Control Systems

1) APU Systems

An auxiliary power unit(APU), shown in Figure 4-2-1, is a compact, self-contained unit that provides electrical power and compressed air during periods of airplane ground activity, or in flight if needed. In some cases, the APU may only be used when the airplane is on the gorund. The unit consists of a small gas turbine engine with engine controls, mountings and enclosures necessary for safe and continuous operation. The APU gas turbine compressor bleed system, or a load compressor, is connected to the airplane's pneumatic system and supplies pneumatic power for main engine starting and other airplane functions. The APU frees the aircraft from dependence on ground power

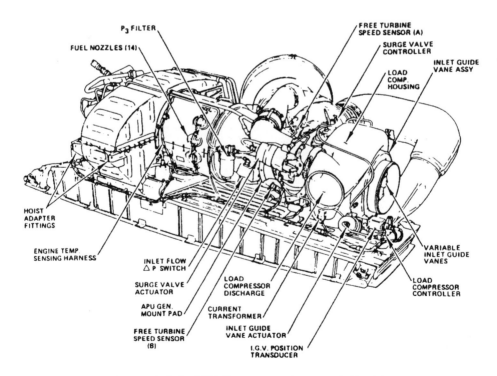

Figure 4-2-1 APU Component Locator-L.H. side

4-2. APU, Pneu, and ECS   175

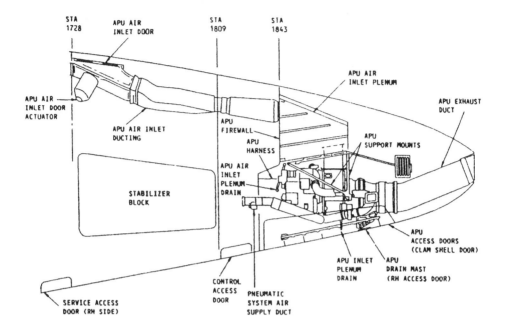

Figure 4-2-2 Typical APU Installation

equipment. The APU generator supplies electrical power to engergize the airplane's electrical systems. Fire detection and extinguishing systems are installed to provide fire protection for the APU.

The APU is installed in the tail cone of most airplanes, isolated from flight critical structure and control surfaces by a firewall, as shown in Figure 4-2-2. The APU provides power for systems operation, generally when the main engines are not running.

A 24-volt battery is provided for APU starting. The main airplane battery switch must be ON to operate the APU. This battery is provided with a battery charger which is disconnected during APU starter engagement. Power for normal charging is provided by the aircraft's electrical system.

Depending upon the aircraft, the APU drives one or two generators that are generally identical to the engine driven generators. The APU generators are usually capable of supplying all or most of the electrical load needed by the airplane. If maximum shaft output power is being used in conjunction with pneumatic power, pneumatic loading will be modulated to maintain a safe APU exhaust gas temperature.

Fuel is normally supplied to the APU from one of the airplane's main tanks. Fuel can be supplied from any tank through the crossfeed system.

The APU oil system is a self-contained system consisting of independent supply, pumps, regulator, cooler, filters, and indicator. Switches on the APU control panels in the cockpit operate the APU. Sometimes the APU can be controlled from a ground accessible control panel.

2) Pneumatic Systems

Compressed air for the pneumatic system can be supplied by the engines, APU, or a high pressure ground air source. The APU or ground source would supply the pneumatic system prior to engine start. The engines supply bleed air for pneumatics after engine start. The following systems normally rely on or are examples of systems that use pneumatics for operation:
- Air conditioning/pressurization
- Wing and engine anti-ice
- Engine cross starting
- Hydraulic reservoir pressure
- Air driven hydraulic pumps

A. Engine Bleed Air

Air is generally bled from an intermediate stage(IP stage) or the low pressure stage(LP stage) of the engine's compressor to provide pressure air for the pneumatic system. At low engine speeds, when the pressure from the IP or LP bleed air is insufficient to meet pneumatic system needs, air is automatically bled from the high pressure(HP) stage bleed.

The transfer from the low pressure bleed to the high pressure bleed is achieved by means of a pneumatically operated high pressure regulating valve at the HP stage outlet as shown in Figure 4-2-3.

Downstream of the junction of the IP and HP ducting, air is admitted into the duct by a pneumatically controlled butterfly valve, which acts as a shutoff and as a pressure regulator valve. The IP check valve prevents reverse flow of the high pressure bleed air from entering the IP bleed stage of the engine.

Cooling air is directed from the fan stage of the engine, through a heat exchanger, to regulate precooling of the hot bleed air from the engine's compressors. When the engine is started and delivery of hot bleed air is initiated, the pre-cooler modulating valve(fan air valve) modulates the flow of fan cooling air through the heat exchanger and regulates the duct air temperature.(See Figure 4-2-5). Some engine bleed systems use an air cleaner in the air bleed system to filter the air used for air conditioning before it enters the pneumatic system.

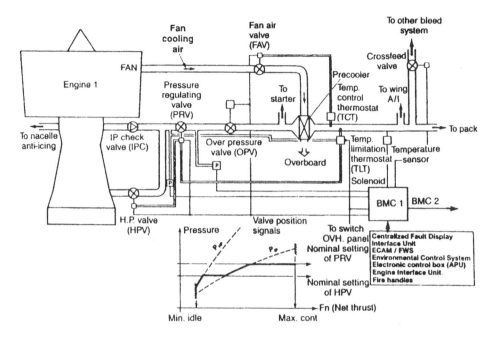

Figure 4-2-3 Airbus A320 bleed system schematic

3) Environmental Control Systems

A large transport aircraft must provide a cabin environment that is comfortable for the crew and passengers. The aircraft operates at altitudes where temperature and pressure extremes would make the cabin unbearably cold and normal breathing would be impossible. On the other hand, the aircraft must also be comfortable when it's on the ground on a hot day. This will require a system of heating and cooling the cabin air and a system for maintaining cabin pressure with enough oxygen to breathe.

The environmental control system's function is to control the cabin temperature and

178 제4장 Aircraft System Description

pressure. It accomplishes this by using two systems: an air conditioning system and a pressurization system. The air conditioning and pressurization systems normally use engine bleed air. Engine bleed air is hot and under pressure, as mentioned earlier. By passing this bleed air through the air conditioning packages(packs), the cabin air temperature is adjusted to the levels called for by the cockpit controls.

The APU bleed air, or ground high pressure air source, may be used to supply high pressure pneumatic air to the packs for normal usage of the air conditioning system on the ground. In some cases, ground conditioned air is used to cool or heat the aircraft cabin on the ground before flight. This air should not be confused with ground high pressure air mentioned previously. Ground conditioned air is conditioned by a unit separate of the aircraft. It does not pass through the packs, passing instead directly into the cabin distribution system. control of the cabin pressure will be discussed later in this chapter.

Air is directed from the pneumatic manifold through the pack valve which controls the air entering the packs. Most aircraft have either two or three air conditioning packs. In Figure 4-2-4, an air conditioning pack is illustrated to show the main components of a pack. Some of these components are the heat exchangers, air cycle machine, ram air cooling doors, water separator, anti-ice valve, and the mixing valve. The terminology for these components might differ from airplane to airplane but these basic components are used in almost every air-conditioning pack.

The heat exchangers are air to air radiators which are used to cool the hot bleed air by passing ambient air through them. The ram air doors control the air flow through the heat exchangers.

The air cycle machine is an air cooling device that is made up of a compressor and an expansion turbine connected together by a common shaft. The air cycle machine changes the hot pneumatic system air to cold air for air conditioning by transforming heat into mechanical energy, as the air is expanded by the turbine.

Water that has condensed out of the air by the operation of the air cycle machine is removed by the water separator. To prevent icing in the water separator, a temperature sensor will signal the water separator anti-ice valve to provide warming air automatically. This will provide a temperature above freezing in the water separator.

The air mixing valve regulates the mixture of cold and hot air for distribution into the passenger and crew cabins. It is a combination of two or more interconnected valves in one body, and is controlled from the temperature control panel. As one valve opens, the other is driven closed, which allows hot or cold air to mix providing the proper temperature.

4-2. APU, Pneu, and ECS   179

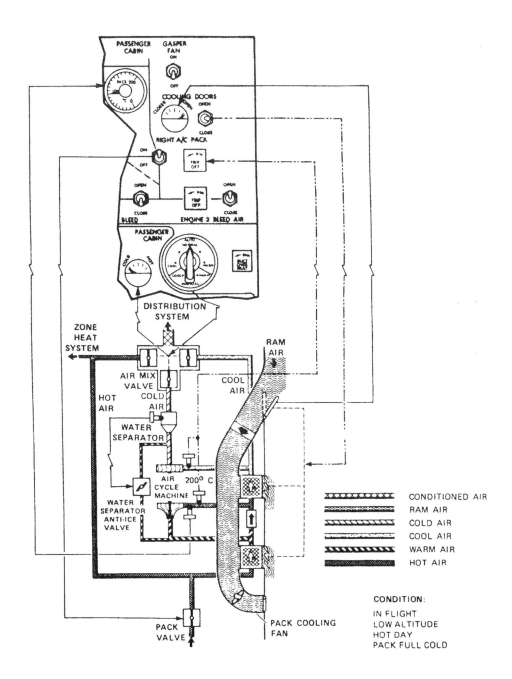

Figure 4-2-4 Boeing 727 air conditioning pack

When the pack valve is turned on, as shown in Figure 4-2-4, air enters the pack and is split, with some of the air going to the hot air mix valve and the rest entering the pack. It passes through the primary heat exchanger where it is cooled to become warm air. It then passes through the compressor of the air cycle machine and is compressed and reheated. The air then passes through the secondary heat exchanger and is cooled again. Part of the air can then go to the cool air mix valve. The other part of the air passes through the air cycle machine expansion turbine where it is cooled even further. The air that has passed through the air cycle machine turbine then passes through the water separator and then on into the cold air mix valve.

By controlling the mixing valves, the temperature called for can be supplied to the distribution system. The temperature sensor in the water separator will control the anti-ice valve to allow enough warm air to mix with the air exiting the expansion turbine to keep the air temperature above freezing.

The cooling door positions are controlled either automatically or manually in the cockpit to adjust the volume of the air passing through the heat exchangers. On the ground, or in slow flight modes, there is generally an auxiliary method of passing air through the heat exchangers, such as an electric pack cooling fan or a fan attached directly to the air cycle machine. Although each aircraft system works somewhat differently, they all use these basic components and principles for their operation.

4) Pressurization systems

engine bleed air is utilized for cabin pressurization. Air from the compressors of the engines, which is pressurized and temperature conditioned by the air conditioning packages(packs), is distributed into the cabin and flight deck as shown in figure 2-5. Air is continually blown into the cabin to maintain proper pressure at altitude.

As mentioned earlier, the air pressure and oxygen level at high altitudes is insufficient to provide a suitable environment for the crew and passengers.

To compensate for this, the amount of pack air that can escape from the cabin area is controlled by the outflow valve as shown in Figure 4-2-5. By closing the outflow valve, the cabin will be pressurized more, thus lowering the cabin altitude. If the outflow valve is opened, it will allow more air to escape, thus increasing cabin altitude. As the pressure in the pressurized area of the aircraft increases, the atmospheric altitude inside the cabin decreases. And conversely, as the cabin pressure decreases, the cabin altitude increases.

4-2. APU, Pneu, and ECS   181

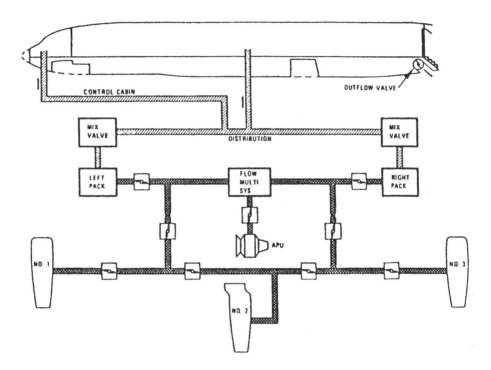

Figure 4-2-5 Pressurized pack air distributed to cabin

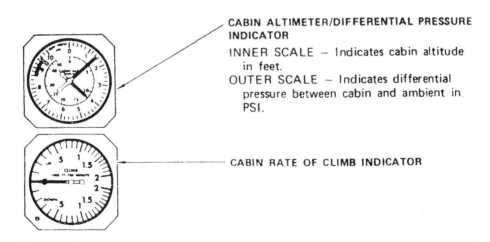

CABIN ALTIMETER/DIFFERENTIAL PRESSURE INDICATOR
INNER SCALE — Indicates cabin altitude in feet.
OUTER SCALE — Indicates differential pressure between cabin and ambient in PSI.

CABIN RATE OF CLIMB INDICATOR

Figure 4-2-6 Pressurization system instruments

Instruments in the cockpit read the cabin altitude, cabin rate of climb, and pressure differential, illustrated in Figure 4-2-6. The cabin altitude(inside pressure)cannot be held at sea level at high altitudes because of the pressure differential between the inside of the cabin and the outside of the cabin. The structure of the airplan is limited by a maximum differential pressure. Most maximum pressure differentials range from about 8.0 PSI to 8.9 PSI. this means the pressure in the aircraft cabin is 8.9 PSI highter than the outside of the cabin. If the pressure inside the aircraft was allowed to build up beyond the maximum pressure differential, the cabin structure could fail.

When the maximum pressure differential is reached, the cabin altitude must be increased(inside pressure decreased). This will cause the cabin altitude to climb, but at a much slower rate and at a much lower altitude than the aircraft. Safety valves(outflow valves) can open if the maximum pressure differential is exceeded and allow air in the cabin to escape. This will decrease the pressure differential and prevent any structural damage from occurring.

The outflow valve position is normally controlled by the system's pressure controller. Most aircraft have an indicator in the cockpit which informs the flight crew of the outflow valve position(see Figure 4-2-7).

Pressurization systems generally have different modes of operation. Most aircraft use either the automatic or manual mode. The automatic mode normally controls the system unless there is a malfunction, then the manual mode is selected and the outflow valve is controlled manually by the flight crew.

The pressurization system also allows the aircraft cabin pressure to climb and descend at a much slower rate than the aircraft is actually climbing or descending. On early pneumatic systems the rate of climb indicator is used to adjust the cabin pressure to a comfortable rate while climbing or descending the cabin. Most new style pressurization systems use electronic controllers and alternating current(AC) and/ or direct current(DC) motors to open or close the outflow valve. Some of the larger aircreaft are equipped with two outflow valves.

A. Boeing 747 Pressurization System

The pneumatic system supplies engine compression bleed air to the air conditioning and pressurization systems. The pneumatic system automatically selects either high or low stage bleed air from the engine compressor and delivers pressure regulated and temperature limited air to the pneumatic duct.

4-2. APU, Pneu, and ECS   183

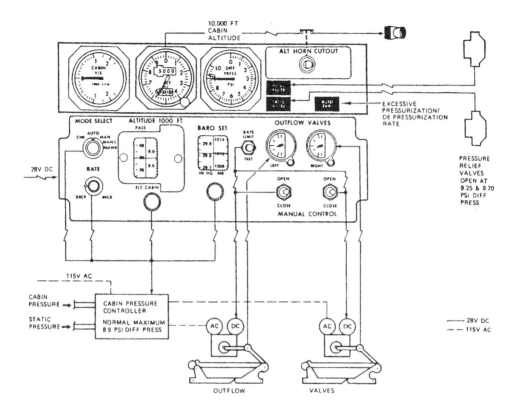

Figure 4-2-7 Boeing 747 pressurization system diagram

Normally during takeoff, climb, cruise, and most holding conditions, the system, draws low stage air.

For descent and other low engine thrust conditions, the high stage is required to provide for system a controlled rate, as selected on the pressurization rate selector.

The cabin pressure selector switch allows selection of two differential pressure operating modes. The 8.9 PSI position provides a maximum differential pressure of 8.9 PSI for all normal high altitude operation.

The 8.0 PSI position provides a differential pressure of 8.0 PSI for low altitude operation. When using the 8.0 PSI position, the reduced differential pressure imposes less stress on the fuselage structure, which provides an increased service life to the airplane.

There is an automatic differential pressure limiter set to 8.0/8.9 PSI which is operative in AUTO and MANUAL modes. Setting a cabin altitude on the flight/cabin altitude selector will result in a differential pressure of 8.0/8.9 PSI when the airplan reaches the corresponding flight altitude shown on the flight/cabin altitude selector(see figure 2-7).

In AUTO and MANUAL modes, the system has a protective feature called rate limit control. This feature is armed when cabin altitude is below 10,000 feet and is activated by cabin vertical speed rates above 3,100 feet per minute ascending or 2,000 feet per minute descending, activation of this feature is indicated by illumination of the RATE LIMIT light.

Negative pressure relief valves are installed in the forward and aft cargo doors. these valves open to prevent external atmospheric pressure from exceeding cabin pressure. The pressure negative relief valves are activated by differential pressure and are activated in any mode of pressurization system operation.

At touchdown, a signal from the landing gear ground safety relay causes the outflow valve to modulate towards OPEN, Depressurizing the cabin at the rate set on the pressurization rate selector. The outflow valves will remain OPEN while the airplane is on the ground.

B. Boeing 747-400 Cabin Pressurization Control system

System mode of operation is by a cabin pressure selector panel which provides automatic and manual modes of operation(see Figure 4-2-8). Output from the panel is supplied to the two cabin pressure controllers(shown in figure 2-9), together with data from the ADCs(air data computers), FMC(flight management computer) and the autopilot system. The controllers alternate control of the system with every flight. The controller not in control will provide back-up control as required.

The rate limit control feature is deactivated and the RATE Limit light is extinguished if cabin altitude exceeds 10,000 feet.

There is an additional protective feature called maximum cabin altitude override which is activated when the cabin altitude exceeds 10,000 feet. This feature, utilizing 115 VAC power(high speed), will drive the outflow valve(s) toward the closed position limiting the cabin altitude between 10,250 and 14,000 feet.

C. Pressurization Safety Relief Valves

Two cabin pressurization relief valves, shown in figure 2-7, are installed to prevent excessive pressure within the airplane. Of the cabin differential pressure reaches 9.25 PSI(or 9.7 PSI as a back-up), one or both valves will modulate OPEN. The operating

4-2. APU, Pneu, and ECS   185

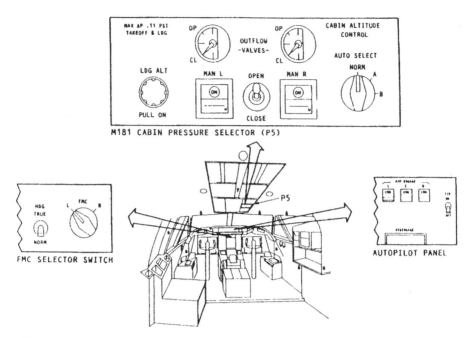

Figure 4-2-8 Cabin pressurization control syystem controls and indications

valve(s) will close when the pressure returns below the activating pressure differential. A PRESS RELIEF light for each relief valve will illuminate whenever its respective valve is open and will extinguish when the valve(s) closes.

The No.2 air conditioning pack will automatically trip to assist in relieving excess cabin pressure if either cabin pressure relief valve opens. The pack cannot be reset until both cabin pressure relief valves are closed.

Thus, controller A will control both the ICU(interface control unit) left and the ICU right during one flight. On the next flight, controller B will control both ICU left and ICU right.

The ICUs integrate controller outputs and provide the operating signals for the two outflow valves. ICU right provides the operating signals to the right outflow valve. ICU left provides the operating signals to the left outflow valve.

The outflow valves can be operated manually from the selector panel through auxiliary panels P212 and P213, bypassing the controllers and interface control units. The outflow valve position is supplied to the interface units, controllers and selector panel. The locations of the pressurization system's major components are shown in figure 2-10.

186 제4장 Aircraft System Description

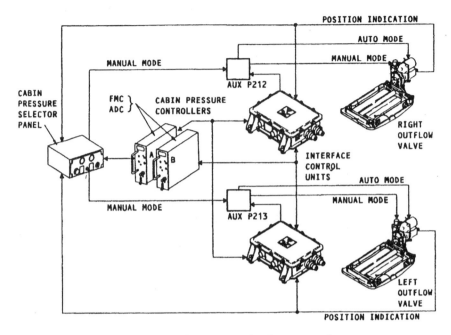

Figure 4-2-9 Cabin pressurization control system

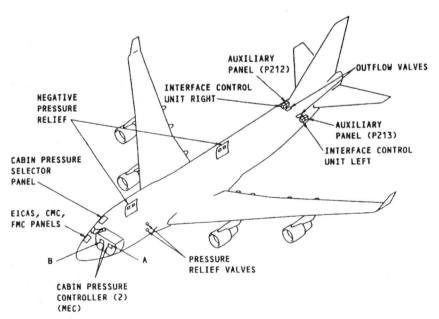

Figure 4-2-10 Cabin pressurization system

4-2. APU, Pneu, and ECS   187

D. Lockheed L-1011

The Pneumatically operated cabin pressurization system, which can be seen in figure 2-11, meters the exhaust of cabin ventilating air to pressurize the entire fuselage area between the forward and aft pressure bulkheads. Areas that are not pressurized are:

- the nose gear wheel well and air conditioning service areas on each side
- the area below the wing center section
- the main landing gear wheel wells and the hydraulic service center between them

Cabin pressure is controlled by a selective, automatic system with manual override control. Major components include the cabin pressure control panel, the cabin pressure controller, the negative pressure relief valve, two safety valves, and two electrically operated cabin outflow valves.

Pressurization control is normally operated in the automatic mode. In this mode, the flight crew selects the desired cabin or aircraft cruise altitude prior to takeoff, and sets

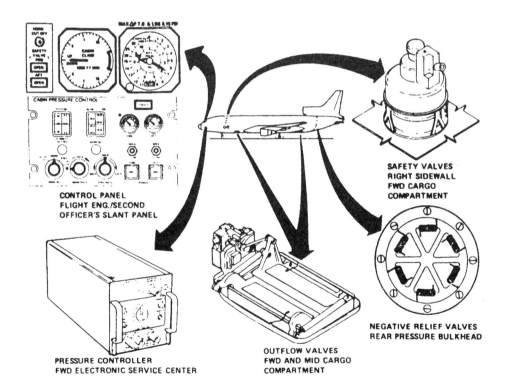

Figure 4-2-11 Cabin pressurization equipment location

barometric correction and landing field altitude prior to descent. The pressurization system is fast responding and pressure transients are controlled below normal threshold of detection.

In the automatic mode, cabin altitude is controlled at a fixed rate of change until the selected altitude is reached. for increasing altitude the rate is 500 feet per minute, and for decreasing altitude it is 300 feet per minute. Other fixed rates of change may be selected by the flight crew with complete automatic operation. The selectable range for increasing altitude is 200 to 1,500 feet per minute and for decreasing altitude the range is 120 to 900 feet per minute. the selectable range for cabin altitude is from 1,000 feet below sea level to 10,000 feet above sea level.

The controller automatically controls the outflow valves to limit cabin pressure differential to a nominal 8.44 PSI. This provides a cabin altitude of 8,000 feet at an aircraft altitude of 42,000 feet. A sea level cabin altitude can be maintained up to an aircraft altitude of 22,000 feet without exceeding the maximum pressure differential.

The outflow valves, located in the bottom of the fuselage, exhaust air overboard at a rate to maintain the desired pressurization schedule. The actuators are driven automatically by AC motors and manually by DC motors to position the valves.

In the manual mode, the outflow valves are controlled by switches located on the selector panel. Manual control of the valves are completely independent of the automatic system. Pressurization control can be maintained in this mode with complete loss of AC power since the DC motor actuators are powered from the stanby battery bus.

E. Safety Relief System

The safety relief system includes both positive and negative pressure relief valves. The two positive pressure cabin safety valves are completely independent, pneumatically operated units, requiring only cabin and ambient pressure references. If the cabin pressure control system fails, these valves will operate to maintain cabin differential pressure below the maximum limit.

The two negative pressure relief valves are simple flapper type valves that open inward. They prevent cabin pressure from becoming less than ambient. This could occur during an extremely rapid descent from altitude. On the ground, the valves open to allow cabin pressure to change with ambient pressure when all other openings in the fuselage(doors, valves, etc.) are closed.

4-3. Anti-icing Systems and Rain Protection

Ice build-up on certain aircraft areas and components can cause serious problems for aircraft operation. It is necessary to prevent ice from accumulating on the aircraft. Ice can form on aircraft surfaces when the outside air temperature(OAT) on the ground and total air temperature(TAT) in flight is 10 degrees C or Below, and visible moisture in any form is present. Visible moisture can be in the form of clouds, fog with visibility of one mile or less, rain, snow sleet and ice crystals. Icing conditions also exist when:the OAT on the ground and for takeoff is 10 drgrees C or below when operating on ramps, taxiways or runways where surface snow, ice, standing water, or slush may be ingested by the engines or feeze on engines, nacelles, or engine sensor probes.

These icing conditions listed here are for example purposes only. The specific icing condition parameters may vary from air carrier to air carrier.

The anti-icing systems on large aircraft generally use hot engine bleed air and electrical heating elements to anti-ice certain areas of the aircraft. Bleed air anti-icing is provided for engines, wings, and sometimes antennas. Operation of the nacelle engine anti-ice and wing anti-ice systems impose a heavy demand on the engine's pneumatic air supply. when bleed air is being used, the engine's power(EPR) is adjusted either manually or automatically to compensate. The parts of the engine that are anti-iced(shown in figure 3-1), may include the inlet guide vanes, the engine nose dome(including Pt2(EPR) Probe), the inlet duct, and the nose cowl leading edges. The areas of the aircraft that are electrically anti-iced can be the air data probes(pitot tubes, static ports, TAT probe), cockpit windows, stall warning, water drain masts, and toilet drain receptacles as shown in Figure 4-3-2. Heating of the cockpit windows is for anti-icing, anti-fogging, and bird impact protection.

Rain protection of the cockpit windows is accomplished normally by windshield wipers and rain repellent. Windshield wipers generally have two speeds along with a park position, which will stow the wipers. The cockpit windows(windshields) can also have a windshield washer system and a fan driven defogger system.

Ice detectors(see Figure 4-3-5) are sometimes used on the outside of the aircraft to warn the pilots that icing is occurring on the aircraft in flight. An annunciator light in the cockpit is generally used to inform the flight crew of icing. The ice detector consists of a probe mounted near the front of the aircraft that vibrates at a set frequency. If ice starts to build up on the probe, the ice will change the probe's frequency and light the cockpit light which will stay illuminated for one minute. This also turns on the probe heater which is connected to a five second timer that will heat up and deice the probe. If the icing condition is continuous, the probe will re-ice and re-energize the timer before the one minute delay expires, and the light will stay on continuously. If the icing condition is temporary, the heater will melt the ice and if no more ice is present, the icing light will go out after one minute.

1) Aircraft Ground Deicing/Anti-icing

Aircraft ground deicing/anti-icing plays a vital role in cold weather procedures to assure that an aircraft is free of ice, frost and snow contamination before takeoff. The equipment most commonly used by airlines for deicing and anti-icing airplanes is the truck-mounted mobile de-icer/anti-icer. These units generally consist of one or more fluid tanks, a heater to bring the fluid to the desired application temperature, an aerial device(boom and basket) to reach the areas that require deicing and/ or anti-icing, and a fluid dispensing system(including pumps, piping and a spray nozzle). The dispensing system is generally capable of supplying the fluid at various pressures and flow rates, and the spray pattern can be adjusted at the nozzle.

There are two basic types of aircraft ground deicing/anti-icing fluids : Type 1(unthickened) Fluids and Type 2(thickened) Fluids. Type 1 fluids have a high glycol content(minimum 80%) and a relatively low viscosity, except at very cold temperatures. Viscosity is a measure of a fluid's ability to flow freely. For example, water has low viscosity, and honey has relatively high viscosity. Because the rate at which fluid flows off of the wing depends, among other things, on the fluid viscosity, a high viscosity fluid is likely to have larger aerodynamic effects than a low viscosity fluid. The viscosity of Type 1 fluids depends only on temperature.

The holdover time, which is the length of time that they will protect the wing from ice, frost and snow, is relatively short for Type 1 fluids. because taxi times are often much longer than the holdover times provided by Type 1 fluids, Type 2(thickened) deicing/anti-icing fluids were developed by fluid manufacturers in cooperation with the airlines. These fluids have significantly longer holdover times than Type1 fluids. Type 2 fluids have a minimum glycol content of 50%, with 45% to 50% water plus thickeners(which increase the viscosity) and inhibitors.

In general, there are two methods for deicing/antiicing an airplane with a mobile unit: a one-step process and a two-step process. The former consists of applying heated fluid onto

4-3. Anti-icing systems and Rain Protection

the airplane surfaces to remove accumulated ice, snow or frost and prevent their subsequent buildup. The primay advantage of this method is that it is quick and uncomplicated, both procedurally and in terms of equipment requirements.

However, in conditions where large deposits of ice and snow must be flushed off of airplane surfaces, the total fluid usage will be greater than for a two-step process, where a more dilute fluid would be used for deicing. Also if Type 2 fluid is used, it may be somewhat degraded if applied at typical deicing temperatures.

The two-step process consists of separate deicing and anti-icing steps. In the deicing step, a diluted fluid, usually heated, is applied to the airplane surfaces to remove accumulated ice, snow or frost. The dilution must be such that protection from refreezing is provided long enough for the second step(anti-icing) to be completed.

During the anti-icing step, a more concentrated fluid(either 100% or diluted appropriately depending upon weather conditions), usually cold, is applied to the

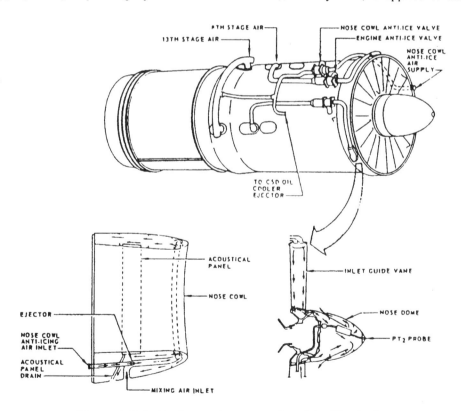

Figure 4-3-1 Engine and nose cowl anti-ice(Boeing)

192 제4장 Aircraft System Description

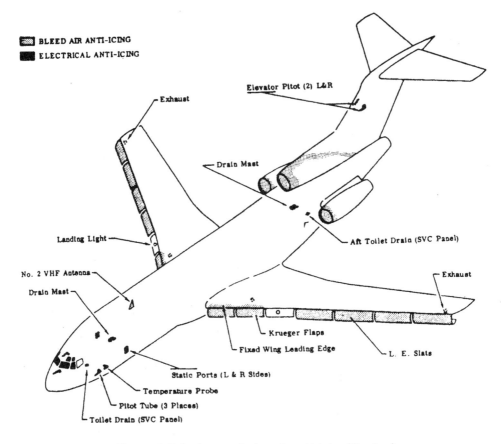

Figure 4-3-2 Areas of aircraft anti-icing(Boeing)

uncontaminated surfaces. Type 1 or Type 2 fluids can be used for both steps, or Type 1 can be used for the first step and Type 2 for the second. This choice would depend upon weather conditions, required holdover time, availability of fluids at a particular station, and equipment capability.

Specific deicing/andti-icing procedures are determined by a combination of common sense and airplane considerations. Maintenance Manuals for each type of airplane provide specific procedures. General precautions include:
- Do not spray deicing/anti-icing fluid directly at or into pitot inlets. TAT probes or static ports.
- Do not spray heated deicing/anti-icing fluid or water directly on cold windows.

4-3. Anti-icing systems and Rain Protection

- Do not spray deicing/anti-icing fluid directly into engine or APU inlets, air scoops, vents or drains.
- Check that ice and/ or snow is not forced into areas around flight controls during ice and snow removal.
- Remove all ice and snow from door and girt bar areas before closing any door.
- cargo doors should be opened only when necessary. Cargo containers should be cleared of ice or snow prior to loading. Apply deicing/anti-icing fluid on pressure relief doors, lower door sills and bottom edges of doors prior to closing doors for flight.
- Do not use hard or sharp tools to scrape or chip ice from an airplane's surface.
- It is not recommended to deice/anti-ice an airplane with the engines or APU operating. Air conditioning pack valves should be closed to prevent fumes from entering the cabin.

2) Boeing 757 Aircraft Anti-icing Systems

A. Wing Anti-ice

Wing anti-ice operation is controlled from the cockpit by a single wing anti-ice switch. Turning on the wing anti-ice switch sends an open signal through the air/ground logic relay to the L and R wing anti-ice valves as can be seen in Figure 4-3-3. The wing anti-ice valves are electrically controlled and pressure actuated. An open wing anti-ice valve permits bleed air to flow to the three leading edge slats, outboard of the engine, on each wing. The air/ground logic prevents wing anti-ice valve operation during ground operation. If a landing is made with the switch on, the wing anti-ice valves automatically close at touchdown. If there is disagreement between the switch and the valve position, an amber VALVE light for each wing illuminates. In the event of valve failure, an EICAS(Engine Indication and Crew Alerting System)advisory message, L or R WING ANTI-ICE, appears.

Due to air being bled from the engine for anti-icing, it is necessary to reduce the maximum EPR(Engine Pressure Ratio). Corrections, to reduce maximum EPR Limits when wing anti-ice is being used, are made automatically. The anti-ice valve position is used to determine the EPR corrections.

An operational check of the valves can be conducted on the ground by pushing the wing anti-ice test switch. The test switch signal bypasses the wing anti-ice switch and opens the valves if air pressure is available. A timed delay automatically closes the valves to protect the wing structure from overheating. The VALVE lights illuminate when the valves are open and extinguish when the valves are closed.

B. Engine Anti-ice

Engine anti-ice operation for the engine cowl is controlled from the cockpit by individual engine anti-ice switches as illustrated in Figure 4-3-4. The engine anit-ice system may be operated on the ground and in flight. To ensure prevention against flame out caused by ice ingestion, engine ignition is automatically activated when the engine anti-ice switch is turned on, provided the engine start selector is in AUTOMATIC. Engine anti-ice must be on during all ground and flight operations when icing conditions exist or are anticipated.

Turning on the engine anti-ice switches sends an open signal to the engine cowl anti-ice valves which permits the cowl leading edge to be anti-iced by engine bleed air. Disagreement between anti-ice switch and the cowl anti-ice valve position illuminates an amber VALVE light on each engine anti-ice switch.

Corrections, to reduce maximum EPR limits when engine anti-ice is being used, are also made automatically. On older aircraft this reduction had to be calculated and set manually. The engine anti-ice switch position is used to determine the EPR corrections. The position of the valves are not considered.

C. Ice Detection

Ice is detected by a sensor on the nose of the airplane. The amber ICING light on the anti-ice panel illuminates and an EICAS advisory message is displayed when ice is detected.

After ice is no longer present, the ICING light is extinguished. The ICING light and EICAS messages are inhibited when the ice detection system is inoperative.

D. Cockpit Window Heat

Cockpit Window Heat The windshields are electrically heated for anti-iceing and anti-defogging. When the cockpit switches are on, power is regulated to the windshields so they are continuously heated. The side windows are electrically heated for anti-fogging only. Overheat protection is provided for the windshields and side windows.

In addition to electrical heating of the windshields. conditioned air is ducted to the top

4-3. Anti-icing systems and Rain Protection 195

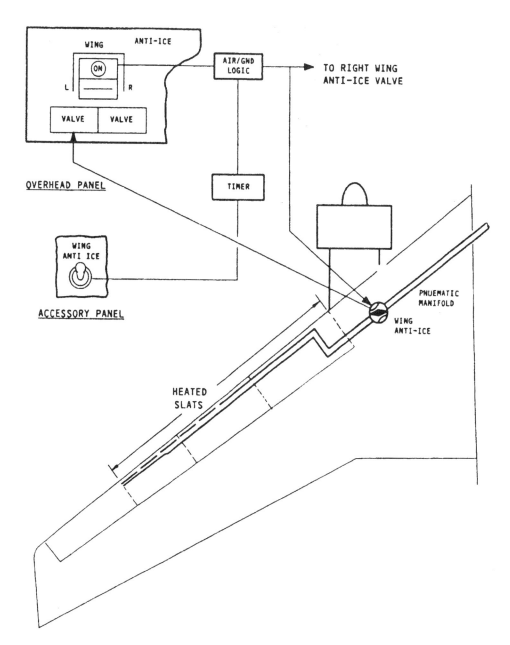

Figure 4-3-3 Boeing 757 wing anti-ice schematic(Boeing)

196 제4장 Aircraft System Description

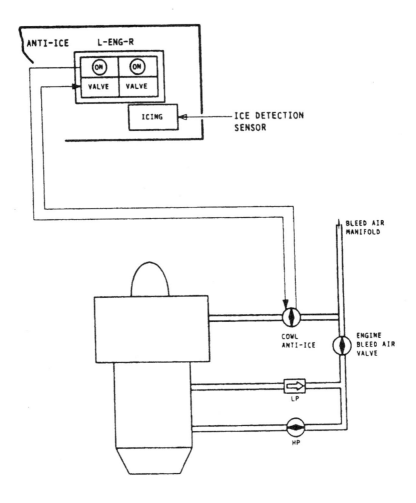

Figure 4-3-4 Boeing 757 engine anti-ice schematic(Boeing)

of the windshield and then flows along the inside surface to provide supplemental anti-fogging. The anti-fogging airflow takes place continuously and is independent of window heat.

Windshield wipers may be used when the flight crew's visibility is impaired by moisture on the windshields. Scratching of the outer windshield surface may occur if the wipers are used on a dry windshield. A three-position windshield wiper selector controls

4-3. Anti-icing systems and Rain Protection

the speed of the electric windshield wipers: LOW and HIGH are two selectable speeds, and OFF stows the wiper in the park position.

E. Rain Repellent

Rain repellent may be used any time rain intensity requires the use of windshield wipers. It should be applied to one windshield at a time to allow the fluid to spread and visibility to improve before application is made to the other windshield. Rain repellent should not be used to clean a windshield. It smears and obscures forward visibility if applied to a dry windshield. If rain repellent is inadvertently applied, do not use the windshield wipers until required for rain removal.

One rain repellent can with a sight gage, pressure gage, and a manual shutoff valve is located in the cockpit. Its application is controlled by individual pushbutton switches on the overhead panel. Each push of the rain repellent switch provides a measured amount of fluid to the associated windshield. Each windshield has independent components to apply the rain repellent, so failure of any component in one system does not affect the operation of the other.

4-4. Electrical Power Systems

The function of the electrical system on a large transport aircraft is to generate, regulate, and distribute electrical power throughout the aircraft. Electrical power is used to operate aircraft flight instruments, essential systems, and passenger services. New-generation aircraft are very dependent on electrical power because of the wide use of electronic flight instrument systems.

Essential power is, just as the name implies, power that is essential to the safe operation of the aircraft. Power for passenger services is provided to light the cabin, to operate the entertainment system, and to prepare food. It is obvious that transport aircraft need a self-contained, dependable, and adequate power generating system.

Most large aircraft use both direct current(DC) and alternating current(AC). Although many different voltages can be used, most large aircraft use 28 volts DC and three phase 115 volts AC at 400Hz to power the aircraft.

By the use of transformer rectifier units(TRU), which change 115 VAC into 28 volts DC, the DC buses and DC components are powered through the aircraft AC generators. Some aircraft also change the 115 VAC to 26 VAC for lighting circuits.

Emergency or standby power can be supplied to the electrical system in the event of complete electrical generator failure from an onboard 28V DC battery. Essential AC power can be obtained during standby power use from a static inverter. The static inverter changes 28V DC to 115 VAC to power essential flight instruments that operate on 115 VAC. Aircraft power is time limited in the standby mode because of battery limits.

1) Power Sources

Some power soruces are engine driven AC generators, auxiliary power units(APU), external power, and ram air driven generators.

Each engine drives an AC generator, shown in Figure 4-4-1, which provides normal inflight power for the entire aircraft. The APU can be used in flight as a backup power source.(One exception is the Boeing 727 aircraft, where the APU can only be operated on the ground.)

External power is used on the ground only with power being provided by a ground power unit(GPU). GPU's can be portable units or stationary units and generally provide AC power through an external plug on the nose area of the aircraft.

A ram air turbine can be used as an emegency source of power. If power from the engine driven generators and the APU is not available in flight, some aircraft incorporate a ram air turbine generator which can be deployed to provide AC power. The aircraft nickel cadmium battery can be used as a power source for standby or emergency power. The battery provides 28 VDC directly and 118 VAC 400Hz by use of the static inverter.

2) System Components

The basic functions of the electrical system's components are to generate power, control electrical power, protect the electrical system, and distribute electrical power throughout the aircraft. The aircraft generators(engine driven and APU) change mechanical energy into electrical energy by using a constant speed drive(CSD) to turn the generator at a set speed. The speed at which the generator turns determines the output frequency of the generator. since most electronic components on the aircraft need 400 cycles per second(Hz), the generator's speed must be held constant. The constant speed drive(CSD), turned by the engine, uses a differential assembly and hydraulic pumps to turn the generator at a constant speed.

The CSD can be mounted on the engine by an external, co-axial, integrated drive generator(IDG), or side-by-side arrangement. The IDG has the CSD drive and generator integrated together as shown in Figure 4-4-1.

Another common method of CSD mounting is where the generator and drive are mounted side by side, which reduces its size and weight. Most CSD's can be disconnected if a problem is experienced during operation. This also disconnects the generator and a CSD cannot be reconnected in flight.

Transformer rectifier units(TRU) are used to change 115VAC, 400Hz to 28 V DC as mentioned earlier. To do this, the TRU must use a transformer to reduce the voltage from 115 volts to 28 vlots. The rectifier then changes the AC to DC current. Generally, each aircraft AC bus will feed a TRU which feeds each DC bus. Both AC and DC currents are used by the aircraft during operation.

Each generator has a voltage regulator or a genenator control unit(GCU) which controls the generator output. Generator circuit protection monitors the various electrical system parameters such as voltage, frequency, overcurrent, undercurrent, and a differential fault. Load controls sense real system load to provide a control signal to the CSD's for frequency control. Current transformers are used for current load sensing and protection of a

200 제4장 Aircraft System Description

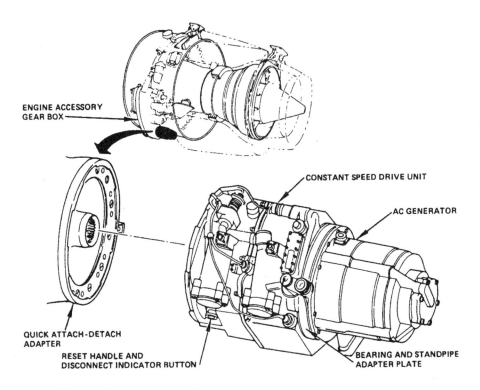

Figure 4-4-1 Integrated drive generator(Lockheed)

differential fault(feeder line short). The electrical system control panel is either located on the pilot's overhead panel or the flight engineer's panel.

3) Electrical System Configurations

A. Parallel Type

There are basically three different electrical system configurations used on large transport aircraft;the parallel, the split bus, and the split system parallel type.

The parallel type is used on the Boeing 727, 747(early series), DC-10, and the L-1011 aircraft. The total aircraft electrical system load is shared equally by all active generators in the parallel system. One advantage of this system is that if one generator fails, the other generators will pick up the load from the failed generator without interrupting primary

4-4. Electrical Power Systems   201

electrical service. The system also has the capability to automatically redistribute the load among the other active generators.

A basic diagram of the Lockheed L-1011's power distribtion system is shown in Figure 4-4-2. The AC tie bus enables paralleled generator operation. Bus tie breakers permit paralleled or isolated generator operation, and prevent the paralleling of external power with operating generators.

Each of the four generators is controlled and protected by its own generator control unit(GCU) which also controls the generator breaker in its channel. Bus tie breakers 1, 2 and 3 are controlled by GCU's 1, 2 and 3 respectively. Generator control current transformers(GCCT's) monitor line current and supply signals to the GCU's for control and protective functions, and to load controllers(LC's) during paralleled generator operation. The load controllers control IDG speed to maintain equal real load division between paralleled generators. Differential protection current transformers(DPCT's) monitor feeder cables for open and shorted conditions.

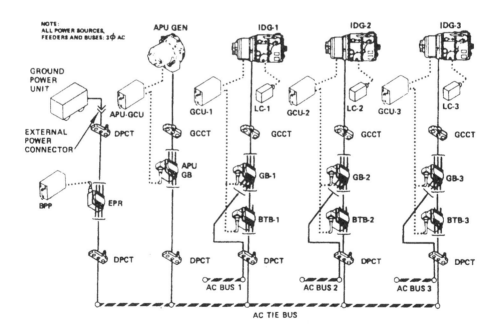

Figure 4-4-2 Basic AC control and distribution system,
Lockheed L-1011(Lockheed)

B. Split Bus System

some examples of aircraft that utilize the split bus electrical system configuration are the Boeing 737, 757,767 and McDonnell Douglas DC-9. In the split bus system, illustrated in Figure 4-4-3, the generators are not operated in parallel. Each generator supplies power separately from the other generator to its respective aircraft bus. The only times the two generator channels are connected is when the aircraft is on external power, APU power, or if one generator fails. The generator channels are electrically connected by bus tie breakers(BTB's) which open and close automatically depending upon the source of power. When external power or APU power is being used to supply power, the BTB's are closed connecting both generator channels togetherWhen an engine driven generator (number 1) comes on line, the number 1 BTB opens as the generator accepts the system load. As the other engine driven generator comes on line, the number 2 BTB opens and now each generator channel is independent in its operation, A more thorough discussion of this type of electrical system configuration is contained later in this chapter.

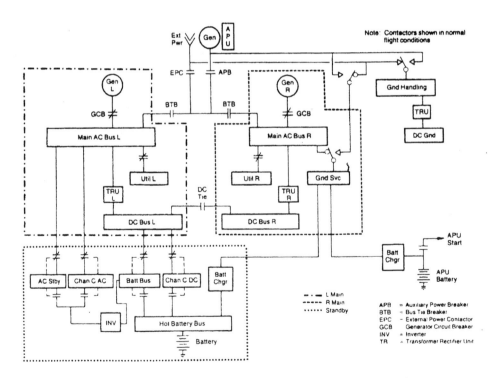

Figure 4-4-3 Boeing split bus electrical system(Boeing)

C. Split Parallel Systems

The split system parallel type electrical system configuration can operate with all generators in parallel or it can be split or electrically disconnected by a split system breaker. The split system breaker, when closed, connects both halves of the synchronous bus. When open, it splits the synchronous bus into two separate halves, or systems. The split system parallel type is used on the Boeing 747-400 aircraft.

4) Boeing 747 Electrical Power System

The 747-400 electrical pwer system, shown in figure 4-4, is an example of a split system parallel type, The systems include:
- External/APU power
- Engine powered generators
- DC power
- Standby power
- Load distribution
- Indicating system

External APU power supplies electrical power during ground operations. It includes two external power sources and two APU generators. Electrical power supplied by the engines is used for all normal flight operations. Engine power consists of four integrated drive generators(IDG). The DC power system supplies those loads requiring DC power. It includes the batteries and transformer rectifier units.

Standby power supplies power to selected loads when the primary power sources have failed. The load distrubution system is used to control and distribute AC and DC power throughout the airplane. The indicating system includes the electrical power interfaces with EICAS and the central maintenance computer and related displays. Electrical power is distributed in the airplane through the AC and DC distribution systems which are illustrated in Fighre 4-5. The AC distribution system is made up of four main buses and several auxiliary buses.

The AC buses are powered by the IDG's, APU generators, or external power carts. An IDG is connected to its respective bus by closing the associated generator circuit breaker(GCB). Parallel operation of the IDG's is accomplished when the associated bus tie breakers(BTB's) and GCB's are closed. The synchronizing(synch) bus is divided into two sections by a split system breaker(SSB). With proper use of the breakers(See Figure 4-4-5),

204 제4장 Aircraft System Description

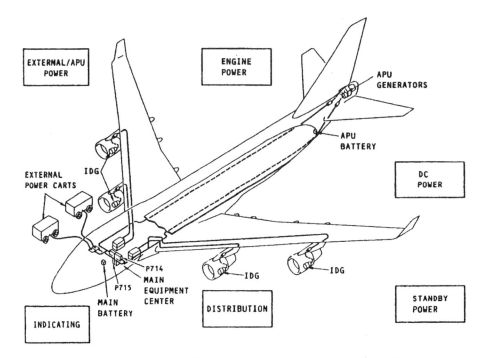

Figure 4-4-4 747-400 Electrical power system(Boeing)

any generator can supply power to any load bus and any combination of the IDG's can be operated in parallel. The power output of the APU Generators or external power carts can also be connected to the load buses by closing the auxiliary power breakers(APB's) or external power contactors(XPC).

There are two transfer buses: the captain's and the first officer's. The captain's transfer bus is normally powered from AC bus 3. If AC bus 3 is not powered, the transfer bus will transfer to AC bus 1 for power.

The first officer's transfer bus is normally powered from AC bus 2, but will transfer to AC bus 1 if AC bus 2 should lose power. The standby AC bus is normally powered, but will switch to the static inverter if the AC bus is not powered.

The ground handling bus is powered from either an APU generator or external power through the ground handling relay(GHR). The bus is powered automatically whenever external power or APU power is available. The ground handling bus is not powered in flight.

4-4. Electrical Power Systems 205

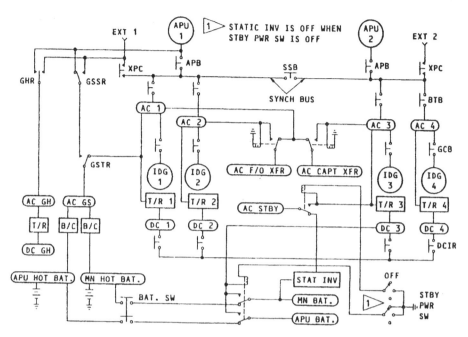

Figure 4-4-5 Boeing 747-400 Electrical power system schematic(Boeing)

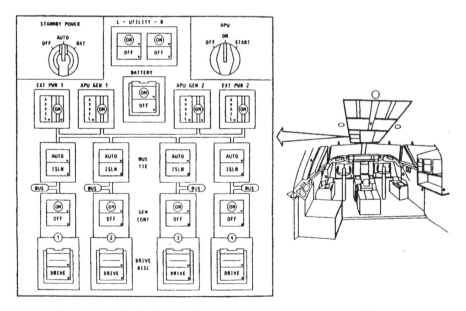

Figure 4-4-6 747-400 Electrical system control panel(Boeing)

Power to DC buses 1,2,3 and 4 is supplied by transformer rectifier units(TRU"s). The DC buses parallel through the DC isolation relays(DCIR). The DC ground handling bus is powered when the AC ground handling bus is powered.

A. External Power Operation

Three-phase, 400Hz, 115 volt AC power is supplied to the airplane by two exteranl power carts through the external power receptacles. Before the power is connected to the airplane systems, two bus control units(BCU's) sample it. The BCU's check for proper seating of the plugs, voltage, frequency, phase rotation and that the plug interlock system is not shorted to the main feeder wires. When the BCU's are satisfied that all of these conditions are met, they turn on the external power AVAIL lights on the electrical system control panel in the flight deck, shown in Figure 4-4-6.

There are two momentary switches labeled EXT PWR 1 and EXT PWR 2 on the electrical system control panel. Pressing EXT PWR 1 signals BCU 1 to close its associated external power contactor(XPC). BCU 2 closes the split system breaker(SSB) at this time if no other power is on the synch bus. Power is now on the airplane and is being supplied by the aft power cord. Pressing EXT PWR 2 signals BCU 2 to trip the SSB and close its associated external power contactor. Now EXT PWR 1 is powering the left side of the synchronizing(synch) bus and EXT PWR 2 is powering the right side of the synch bus.

B. APU Power

Three phase 400 Hz AC power is supplied to the airplane by the APU driven generators. Before the power is connected to the airplane systems, two auxiliary generator control units(AGCU) monitor the power to ensure that it is at the proper voltage and frequency. When the AGCU's are satisfied that the power requirements are met, they signal the BCU's to turn on the two APU generator AVAIL lights on the electrical system control panel in the aircraft to be powered by the APU generators in a similar manner as external power.

C. Integrated Drive Generator

The IDG portion of the engine power system, shown in Figure 4-4-7, involves the mechanical aspects of the IDG and its oil cooling components. The constant speed drive(CSD) portion of the IDG is a hydromechanical device. It adds or subtracts speed from the variable input of the engine gearbox to maintain the IDG generator at 12,000 RPM.

4-4. Electrical Power Systems   207

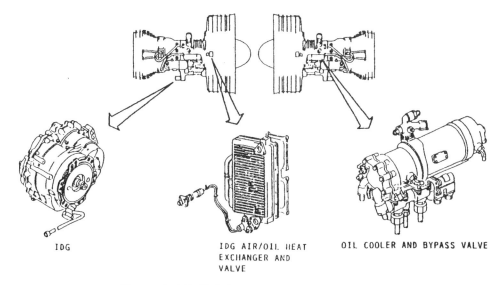

Figure 4-4-7 IDG System components(Boeing)

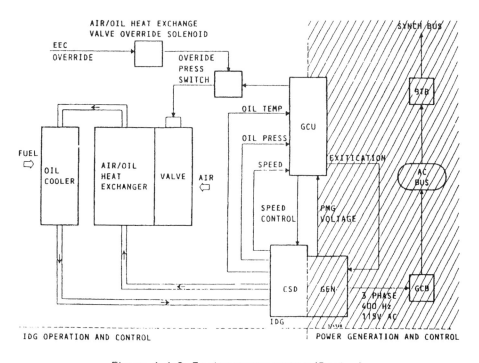

Figure 4-4-8 Engine power system(Boeing)

IDG oil is used for speed control, lubrication, and cooling. Heat generated by the IDG is cooled by passing the oil through an air/oil heat exchanger and a fuel cooled oil cooler. The cooled oil is then returned to the IDG. The oil cooler is the primary means of cooling the oil but is assisted by the air/oil heat exchanger under certain operatin conditions. Four generator control units(GCU's) are used to provide control, protection, and built-in test equipment(BITE) for their respective generator channels.

The GCU receives oil temperature, oil pressure, and IDG speed signals as can be seen in Figure 4-4-8.

It sends speed commands to the IDG governor and control signals to the air/oil heat exchanger valve.

Speed commands from the GCU to the IDG are used to control IDG speed to maintain a reference frequency. Air/oil heat exchanger valve control signals are sent to the valve to control its position. BITE circuitry is provided in the GCU to identify a malfunction or failure of the electrical system. The BITE mode of operation is initiated under two conditions: anytime a protective trip occurs and periodically(automatically).

· DC power
· Standby power
· Load distribution
· Indicating system

4-5. Flight Control Systems

1) Boeing 727

Large aircraft control systems are very similar to small aircraft in that they control the aircraft around three axes. The longitudinal, vertical, and lateral axes use controls which allow movement of the aircraft so it can revolve to change flight directions.

Motion about the longitudinal axis, which runs through the nose and out the tail parallel to the fuselage, is the axis that produces roll.

Motion about the lateral axis, which is wing tip to wing tip, produces pitch.

Movement of the aircraft around the vertical axis produces yaw.

Flight controls that control the aircraft about these axes are the primary flight controls which consist of the ailerons(longitudinal axis, roll), the elevators(lateral axis, pitch), and the rudder(vertical axis, yaw). The primary flight controls for a Boeing 727 are shown in Figure 4-5-1.

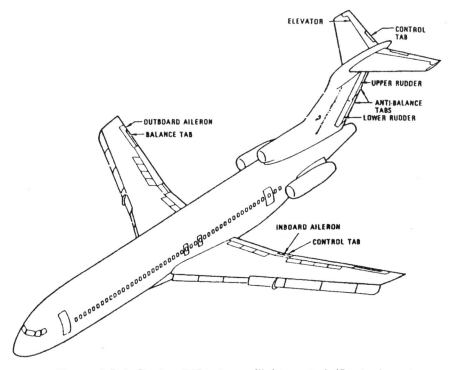

Figure 4-5-1 Boeing 727 primary flight controls(Boeing)

210 제4장 Aircraft System Description

Most large aircraft's primary flight contrl surfaces are hydraulically powered. The hydraulic power sources are divided between multiple aircraft hydraulic systems to minimize the impact of the loss of any one of the aircraft's hydraulic systems.

Each hydraulic system uses its own individual actuating cylinders, in case of a hydraulic fluid leak in that system. Each control surface uses two or more hydraulic systems with each system having individual actuating cylinders connected to the flight controls.

Primary pitch control is provided by elevators mounted on the trailing edge of the horizontal stabilizer. Elevator movement is normally controlled by the elevator power control units. Each unit on the Boeing 727 is powered by both hydraulic systems as shown in Figure 4-5-2. If one system fails, either system will power both elevators. If both

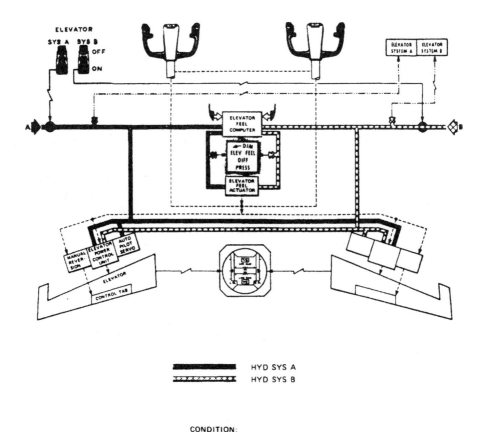

Figure 4-5-2 Pitch control elevator system, Boeing 727(Boeing)

4-5. Flight control Systems   211

hydraulic systems are lost, the control tabs unlock and allow mechanical control of the elevators.

Pitch trim is sometimes provided by moving the leading edge of the horizontal stabilizer up or down. The Boeing 727's horizontal stabilizer is operated electrically by either a main trim motor or an autopilot/cruise trim motor as shown in Figure 4-5-3. A mechanical stabilizer brake will stop stabilizer movement any time control column movement is in the opposite direction to the trim wheel rotation. If the trim motors fail, the

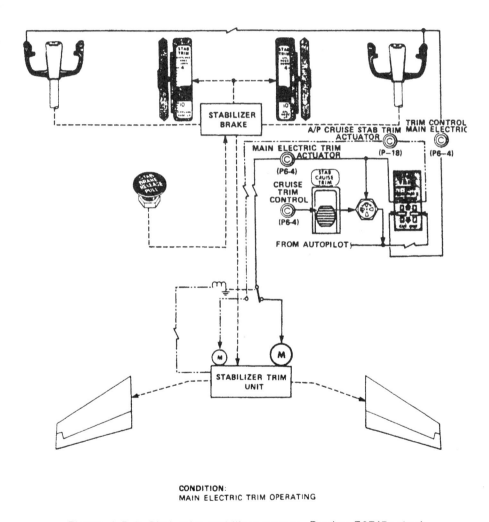

CONDITION:
MAIN ELECTRIC TRIM OPERATING

Figure 4-5-3 Pitch trim stabilizer system, Boeing 727(Boeing)

212 제4장 Aircraft System Description

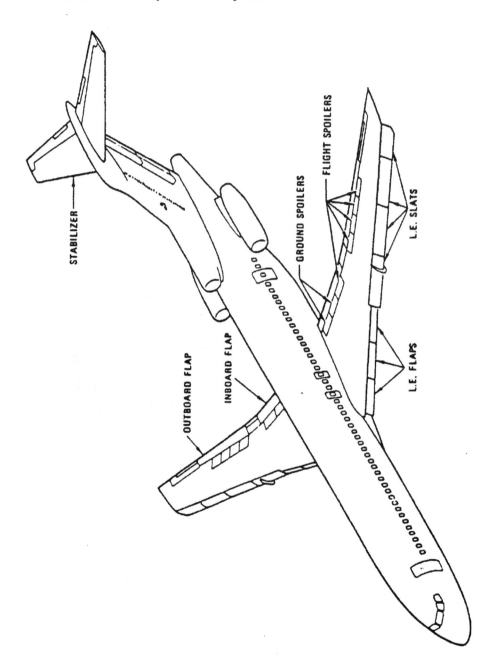

Figure 4-5-4 Secondary flight controls, Boeing 727(Boeing)

4-5. Flight control Systems

stabilizer trim can be operated manually.

Yaw control is provided by the rudder or rudders generally mounted on the trailing edge of the vertical stabilizer of fin. The Boeing 727 yaw control system, uses two full time yaw dampers to move the rudders to oppose airplane yaw.

Yaw damper operation is controlled by yaw damper rate gyros which augment rudder pedal inputs. Sometimes anti-balance tabs are used on each rudder which move in the direction of rudder displacement, providing the effect of a larger rudder. Rudder pedal feel is supplied by a feel and centering mechanism. The lower ruder can be operated by the standby hydraulic system in the event of normal hydraulic pressure loss.

If hydraulic pressure is lost to the primary controls, the 727 aircraft uses manual reversion which incorporates control tabs mounted on the ailerons and on the elevator. These controls are unlocked from their actuating cylinders and manual inputs are provided to the control tabs from the cockpit.

The wings on large transport aircraft are generally swept back at an angle of about 30 to 35 degrees to reduce high speed drag(see Figure 4-5-1). The lateral control of the aircraft involves creation of differential lift between the two wings which is accomplished on most aircraft by the use of ailerons mounted on the wing's trailing edge.

Some large aircraft use two sets of ailerons, an inboard and outboard set. As the aircraft's speed is increased, the aerodynamic loads on the ailerons tend to twist the wings because the wing is more flexible near the tip.

To overcome this problem, some aircraft use the technique of locking out the outboard ailerons during high speed flight. The outboard ailerons are locked in the faired(neutral) position when the trailing edge flaps are fully retracted.

Secondary flight controls, shown in figure 5-4, are used to aid primary controls, relieve control pressure, and increse or decrase the wing's lift. Some secondary controls are spoilers, wing flaps(trailing edge), leading edge devices(slats), and control trim systems.

Spoilers can be used in conjunction with the ailerons to create a wing lift differential. Spoilers are installed on the upper wing surfaces dircetly forward of the flaps and consist of panels that assist the ailerons when maximum roll rates are required. Spoilers reduce the lift on the wing which moves down as the aircraft rolls(banks). Aileron control is supplemented by the flight spoilers which are controlled by the spoiler mixer. The aileron and spoiler system for the Boeing 727 aircraft is illustrated in Figure 4-5-5.

The spoiler mixer senses aileron movement and automatically provides the correct amount of spoiler deflection. Spoilers can also be used as a backup if aileron control is lost.

214 제4장 Aircraft System Description

Spoilers, when used as landing speed brakes, extend upward about 60 degrees during landing, which reduces the wing's lift. This increases drag with a resultant increase in weight on the wheels of the aircraft. This can shorten the landing roll by increasing the brake effectiveness.

Flight spoilers or speed brakes can also be used in flight to reduce the aircraft's speed. Ground spoilers, as the name implies, can only be used on the ground after landing to decrease lift and increase drag.

Wing flaps are employed to increase lift for takeoff and landing operations. Generally two pairs of flaps are located on the wing trailing edge, an inboard and an outboard pair.

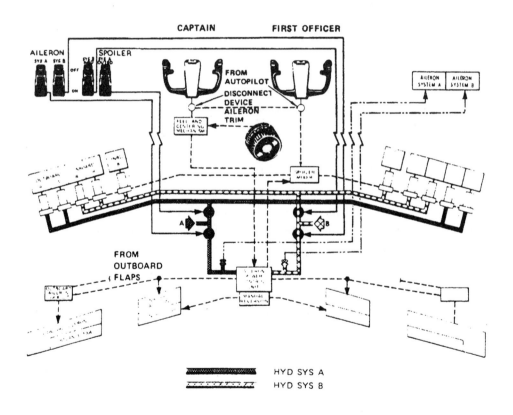

Figure 4-5-5 Roll control aileron system, Boeing 727(Boeing)

4-5. Flight control Systems   215

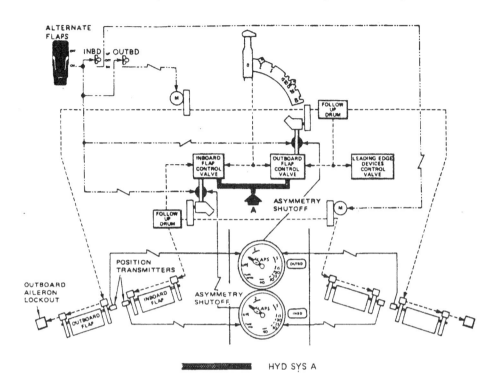

CONDITION:
FLAPS RETRACTING

Figure 4-5-6 High lift devices, trailing edge flaps, Boeing 727(Boeing)

On many aircraft the flaps are powered by the hydraulic system for normal operation with electric motors used for alternate operation.

On the Boeing 727 airplane two pairs of triple-slotted trailing edge flaps, outborad and inboard, are normally operated by hydraulic pressure. The 727's flap position determines the availability of the outboard ailerons and controls the normal operation of the leading edge devices. The flap system for the 727 can be seen in Figure 4-5-6.

Leading edge devices are installed on the wing's leading edge to increase lift during takeoff and landing. Leading edge devices on the 727 consist of slats or flaps and are normally operated by hydraulic pressure.

216 제4장 Aircraft System Description

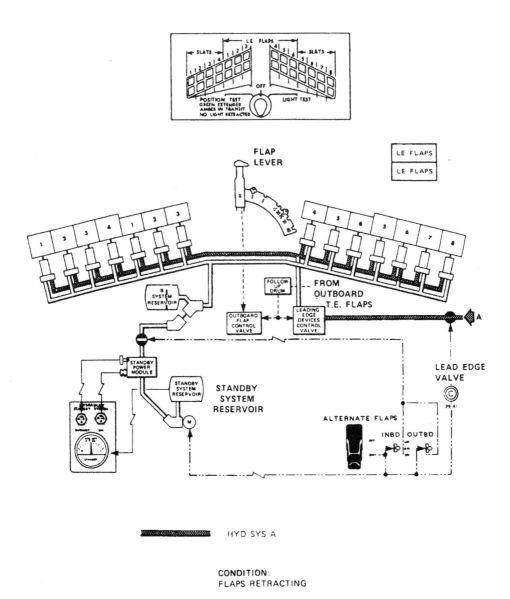

Figure 4-5-7 High lift devices, leading edge devices, Boeing 727(Boeing)

4-5. Flight control Systems

The leading edge devices(flaps and slats) for a 727, shown in Figure 4-5-7, are controlled by the outboard trailing edge flap position. When the trailing edge flaps reach position 2, two leading edge slats on each wing are extended. With the flaps in positions 5 or greater, all leading edge devices are fully extended. When the flaps are retracted, this sequence is reversed.

A back-up or standby system is generally provided for operation of the leading edge devices if the primary means of operation is lost. Trim systems are provided for most aircraft primary controls by changing the control system's power unit input.

The control feel system provides a variable control force which closely simulates forces a pilot would feel if the controls were operating with only aerodynamic forces on the control surfaces.

The 727 aircraft system uses two major components, the feel computer and the feel control unit. The computer determines the correct control force from airspeed and stabilizer position data which provides a variable hydraulic force to the elevator feel unit.

218 제4장 Aircraft System Description

4-6 Fuel Systems

1) Turbine Engine Fuels

Large transport aircraft use turbofan gas turbine engines, as mentioned earlier, which are designed to operate on a distillate fuel, commonly called jet fuel. Jet fuels are also composed of hydrocarbons with a little more carbon and usually a higher sulphur content than gasoline. Inhibitors may be added to reduce corrosion and oxidation. Anti-icing additives are also being blended to prevent fuel icing.

Two types of jet fuel in common use today are : (1) Kerosene grade turbine fuel, now named Jet A ; and (2) a blend of gasoline and kerosene fractions, designated Jet B. There is a third type, called Jet A-1, made for operation at extremely low temperatures.

Both Jet A and Jet B fuels are blends of heavy distillates and tend to absorb water. The specific gravity of jet fuels, especially kerosene, is closer to water than is aviation gasoline; thus, any water introduced into the fuel, either through refueling or condensation, will take an appreciable time to settle out. At high altitudes, where low temperatures are encountered, water droplets combine with the fuel to form a frozen substance referred to as gel. The mass of gel, or icing, that may be generated from moisture held in suspension in jet fuel can be much grater than in gasoline.

Because jet fuels are not dyed, there is no on-sight identification for them. They range in color from a colorless liquid to a straw-colored(amber) liquid, depending on age or the crude petroleum source. Jet fuel numbers are type numbers and have no relation to the fuel's performance in the aircraft engine.

2) Fuel System Contamination

There are several forms of contamination in aviation fuel. The higher the viscosity of the fuel, the greater is its ability to hold contaminants in suspension. For this reason, jet fuels having a high viscosity are more susceptible to contamination than aviation gasoline. The principle contaminants that reduce the quality of turbine fuels are other petroleum products, water, rust or scale, and dirt

A. Water
Water can be present in the fuel in two forms : (1) Dissolved in the fuel or (2) entrained

4-6. Fuel Systems

or suspended in the fuel. Entrained water can be detected with the naked eye. the finely devided droplets reflect light and is high concentrations give the fuel a dull, hazy, or cloudy appearance. Particles of entrained water may unite to form droplets of free water.

Fuel can be cloudy for a number of reason. If the fuel is cloudy and the cloud disappears at the bottom, air is present. A cloud usually indicates a water-in-fuel suspension. Free water can cause icing of the aircraft fuel system, usually in the aircraft boost pump screens and low pressure filters. Large amount of water can cause engine stoppage.

B. Microbial Growth

Microbial growth is produced by various forms of micro-organisms that live and multiply in the water interfaces of jet fuels. These organisms may form a slime similar in appearance to the deposits found in stagnant water. The color of his slime growth may be red, brown, gray, or black. If not properly controlled by frequent removal of free water, the growth of these organisms can become extensive. The organisms feed on the hydrocarbons that are found in fuel, but they need free water in order to multiply. The buildup of micro-organisms not only can interfere with fuel flow and quantity indication, but more important, it can start electrolytic corrosive action

C. Contamination Detection

Coarse contamination can be detected visually. The major criterion for contamination detection is that the fuel be clean, bright, and contain no perceptible. free water. Clean means the absence of any readily visible sediment or entrained water. Bright refers to the shiny appearance of clear, dry fuels, Free water is indicated by a cloud haze, or a water slug. A cloud may or may not be present when the fuel is saturated with water. perfectly clear fuel can contain as much as three times the volume of water considered to be tolerable. Since fuel drained from tank sumps may have been cold-soaked, it should be realized that no method of water detection can be realized that no method of water detection can be accurate while the fuel entrained water detection can be accurate while the fuel entrained water is frozen into ice crystals.

There is a good chance that water will not be drained or detected if the sumps are drained while the fuel is below 32 degrees F. after being cooled in flight. Draining will be more effective if it is done after the fuel has been undisturbed for a period of time, during which the free water can precipitate and settle to the drain point. The benefits of a settling period will be lost, however, unless the accumulated water is removed from the drains before the fuel is disturbed by internal pumps.

3) Fuel Systems

The aircraft fuel systems stores fuel and delivers the proper amount of clean fuel at the right pressure to meet the demands of the engine or engines. A well designed fuel system ensures positive and reliable fuel flow throughout all phases of flight, which include changes in altitude, violent maneuvers and sudden acceleration and deceleration. Such indicators as fuel pressure, fuel flow, warning signal, and tank quantity are provided to give continuous indications of how the system is functioning.

4) Fuel System Components and Subsystems

The basic components of a transport air craft fuel system include tanks, links, lines, valves, flow indicators, filters, fuel quantity, and misc. warning components. The fuel systems of large aircraft can also be subdivided into several subsystems, which have provisions for fuel jettison(dumping), fuel heating, cross feeding fuel, tank ventilation, and central refueling.

A. Fuel Tank

The location, size, shape, and construction of fuel tank vary with the type and intended use of the aircraft. In most large aircraft, the fuel tanks are integral with the wing or other structural portions of the aircraft. Since integral cell are built into the wings of the aircraft structure, they are not removable. An integral cell is a part of the aircraft structure, which has been so built that after the seams, structural fasteners, and access door have been properly sealed, the cell will hold fuel without leaking. This type of construction is usually referred to as a wet wing.

Each tank is vented to the outside air through a ventilation system, in order to maintain atmospheric pressure within the fuel tanks. In order to permit rapid changes in internal air pressure, the size of the vent is proportional to the size of the tank. Most tanks are fitted with internal baffles to resist fuel surging caused by changes in the attitude of the aircraft. Usually an expansion space is provided in fuel tanks to allow for an increase in fuel volume due to expansion

4-6. Fuel Systems   221

B. Fuel Lines and Fittings

In an aircraft fuel system, the various tanks and other components are usually joined together by fuel lines made of metal tubing. Where flexibility is necessary. lengths of flexible hose are used. The metal tubing is usually made of aluminum alloy, and the flexible hose is made of synthetic rubber or Teflon®. the diameter of the tubing is governed by the fuel flow requirements of the engine.

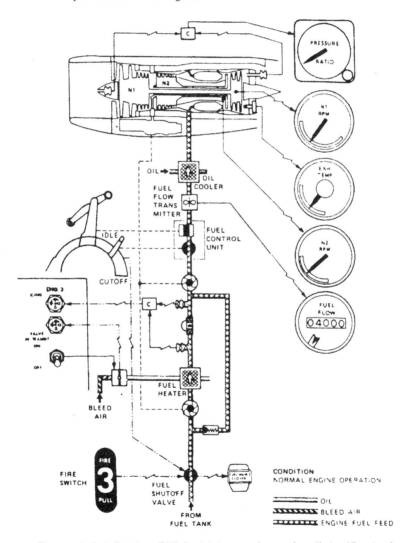

Figure 4-6-1 Boeing 727 fuel icing and warning light (Boeing)

222 제4장 Aircraft System Description

Strainers or filters are used in the fuel system to trap water and other contaminants. Many filters incorporate a bypass in the event that the filter becomes completely or partially closed. If ice is mixed with fuel, the ice can collect in the filter until it begins to stop the flow of fuel. when this happens, a differential pressure switch will illuminate a warning light i the cockpit. This warning light will alert the flight crew to the situation as shown in Figure 4-6-1. By using fuel heat(a process of heating the fuel with engine bleed air), the ice crystals can be melted and the water will be consumed through the engine. If the blockage in the filter was not ice crystals, then the filter blockage is from fuel contamination and use of fuel heat will not correct the problem.

C. Auxiliary Fuel Pumps

The electrically driven centrifugal booster pump, shown in Figure 4-6-2, supplies fuel under pressure to the inlet of the engine-driven fuel pump. This type of pump is an essential part of the fuel system, particularly at high altitudes, to keep pressure on the suction side of the engine-driven pump from becoming too low. Most transport aircraft use two or more boost pumps per fuel tank. Each boost pump generally has an associated low pressure light that illuminates when the pump is switched off. This is illustrated in Figure 4-6-3

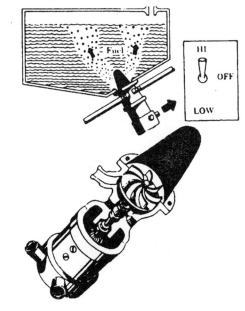

Figure 4-6-2 Centrifugal fuel booster pump

D. Engine-Driven Fuel Pump

The purpose of the engine-driven fuel pump is to deliver a continuous supply of fuel at the proper pressure at all times during engine operation. The type of pump most used is the positive displacement, rotary vane type pump.

A schematic diagram of a typical engine-driven pump(vane type) is shown is figure 4-6-4. Regardless of variations in design, the operating principle of all vane type pump is the same.

4-6. Fuel Systems   223

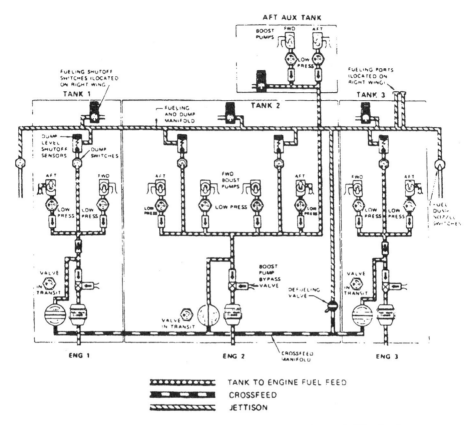

Figure 4-6-3 Boeing 727 fuel system schematic (Boeing)

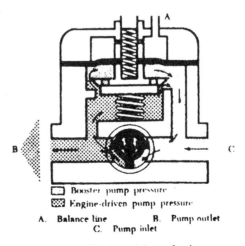

A. Balance line   B. Pump outlet
C. Pump inlet

Figure 4-6-4 Engine-driven fuel pump

224 제4장 Aircraft System Description

E. Valves

Selector valves are installed in the fuel system to provide a means for shutting off the fuel flow, for tank and engine selection, crossfeed, and for fuel transfer. The size and number of ports openings vary with the type of installation.

Selector valves may be operated either manually or electrically. Electrically operated valves have an actuator, or motor, which opens or closes the valve.

F. Fuel Quantity Indicators

Large transport aircraft generally use an electronic type(capacitance) fuel quantity indicating system. This system differs from smaller aircraft types, in that it has no movables devices in the fuel tank. The dielectric qualities of fuel and air furnish a measurement of fuel quantity. Essentially, the tank transmitter is a simple electric condenser. The dielectric(or nonconducting material) of the condenser is fuel and the conducting material is air(vapor) above the fuel. The capacitance of the tank unit at any one time will depend on the existing proportion of fuel and vapors(air) in the tank.

G. Fuel Flowmeter

The fuel flow transmitter used with many turbine engines is the mass flow type, having a range of 500 to 2,500 pounds per hour. It consists of two cylinders placed in the fuel stream so that the direction oh fuel flow is parallel to the axis of the cylinders(see Figure 4-6-5). The cylinders have small vanes in the outer periphery. The upstream cylinder, called the impeller, is driven at a constant angular velocity by the power supply. This velocity to the turbine(the downstream cylinder), causing the turbine to rotate until a restraining spring force balances the force due to the angular momentum of the fuel. The deflection of the turbine positions a magnet in the second harmonic transmitter to a position corresponding to the fuel flow. the turbine position is transmitted to the flight station indicator by means of a selsyn system.

H. Fuel Pressure Gauge

On aircraft where the fuel pressure gauge is located some distance from the engine, a transmitter is usually installed. The pressure transmitter may be a simple cast metal cell that is divided into two chambers by a flexible diaphragm or by electrical transmitters which register fuel pressure on the gauge.

4-6. Fuel Systems

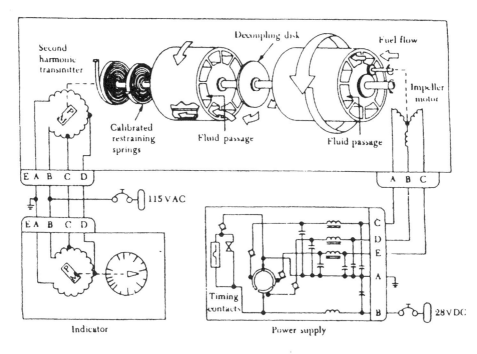

Figure 4-6-5 Schematic of a turbine engine fuel flow indicating system

I. Pressure Warning Signal

In an aircraft with several tanks, there is always the possible danger of allowing the fuel supply in one tank to become exhausted. To prevent this, pressure warning lights are installed in the aircraft as shown in figure 4-6-3.

Normal fuel pressure against the power surface of the diaphragm hold the electrical contact apart. When the fuel pressure drops below specified limits, the contacts close and the warning light turned on. This alerts the flight crew to take whatever action is necessary to boost the fuel pressure.

J. Valve-In-Transit Indicator Lights

On large transport aircraft of the fuel crossfeed and line valves may be provided with a valve-in-transit indicator lights. This light is on only during the time the valve is in motion and is off movement is complete.

K. Fuel Temperature Indicator

A means for checking the temperature of the fuel in the tank and at the engine is provided on most large turbine powered aircraft. During extreme cold, especially at altitude, the gauge cat be checked to determine when fuel temperatures are approaching those at which there may be danger of ice crystals forming in the fuel.

L. Crossfeed System

The main feature of the crossfeed system, shown in figure 4-6-3, is its fuel manifold. As shown, fuel is being supplies from the main tank directly to the engines. the crossfeed valves can be set so that all tanks feed into the fuel manifold and each engine receives its fuel supply from this line.

The auxiliary fuel supply can be delivered to the engines only through the manifold. The main advantage of this system is its flexibility. Should an engine fail, its fuel is immediately available to the other engines, if a tank is damaged, the corresponding engine can be supplied with fuel from the manifold.

M. Central Refueling Systems

An advantage of the central refueling system is that all fuel tank can be refueled at the same time through a single line manifold connection. This method of refueling has greatly reduced servicing time on large aircraft because fuel can be introduced into the fueling manifold under high pressure to fill each tank to the proper level.

N. Fuel Jettison Systems

A fuel jettison(dump) system, illustrated in Figure 6-4-3, is required for transport category aircraft if the maximum take off weight exceeds the maximum landing weight. The maximum take off and landing weights are design specifications and may be found in the Aircraft Type Certificate data sheets.

A fuel jettison system must be able to jettison enough fuel within 15 minutes for transport category aircraft to meet the requirements of the Federal Aviation Regulations. It must be operable under conditions encountered during all aperations of the aircraft.

Design requirements are that fuel jettisoning must be stopped with a minimum of fuel remaining on a turbine powered aircraft for take off and landing and 45 minutes cruging time.

4-6. Fuel Systems   227

5) Boeing 747-400 Fuel System

Fuel storage tanks, shown in Figure 4-6-6, are used to store all fuel within vented areas of the wing, the wing center section and the horizontal stabilizer.

The fuel storage areas are divided into two reserve tanks, four main tanks, the center wing tank, and the horizontal stabilizer tank(HST). These tank sections are integral units using the sealed structure of the airplane to store fuel and vent air.

Each wing has additional surge tank located at the tips. These tanks are not used to store fuel but act as containers for overflow from the fuel tanks. The horizontal stabilizer also has a surge tank, but only in the right wing section. The wing center section is both a primary wing structure and a fuel storage tank. Spanwise beams divide the entire box section into five compartments for structural seasons. The compartments are not sealed or fluid tight from each other. The entire upper and front spar surfaces of the wing center box are coated to provide secondary fuel barrier.

Fuel is supplied to the airplane through the fueling receptacles in either wing leading edge, as illustrated in Figure 4-6-7. The fueling system is controlled with the fueling control panel located in the left wing leading edge only. Electrical power for the system is automatically connected when the fueling panel is opened.

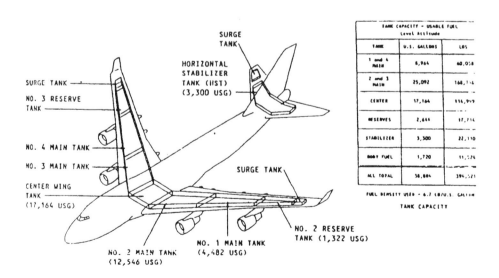

Figure 4-6-6 Boeing 747-400 fuel system schematic

228 제4장 Aircraft System Description

Fuel is supplied to the airplane from an external source with hoses that connect to any of the four fueling receptacles. Once connected, the fueling system is pressurized by opening the manual shutoff valves located at each fueling receptacle. The maximum allowable fueling pressure is 65 PSI.

Once fuel pressure is applied to the fueling system and the power supply selected, three types of fueling control are available : manual, top-off, and preselect. Manual fueling is accomplished by opening the desired refuel valve control switch. Once the desired fuel level is reached, the refuel valve control switch is closed.

Top-off fueling is similar to manual fueling except that the closing of the refuel valve is controled by the fuel quantity indication system(FQIS) processor. When a fuel thank is full, processor closes the refuel valve.

Preselect fueling is also controlled by the FQIS processor. The desired fuel level is selected with the refuel select quantity switches and stored into the FQIS processor memory with preselect switch. The FQIS processor closed the refuel valve for each tank according to the preselected fuel quantity and a programmed loading schedule in the FQIS processor software

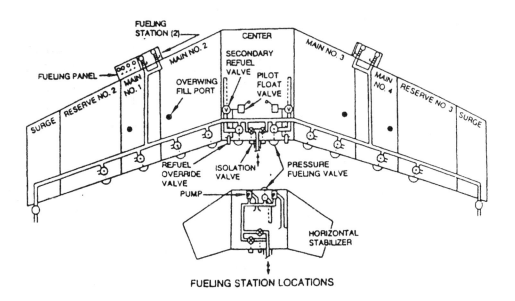

Figure 4-6-7 Boeing 747-400 fuel system(Boeing)

A. Fuel Quantity Indicating System

The fuel quantity indicating system(FQIS). Shown in Figure 4-6-8, computers the quantity of fuel contained in each airplane tank. The calculating function is performed by the fuel quantity processor.

Inputs to the processor come from the fuel sensors in each tank. Fuel quantity calculations for the horizontal stabilizer tank(HST) are performed by the remote electronics unit(REU). The quantity information is transmitted to FQIS processor.

The fuel quantity information is presented on the engine indicating and crew alerting system(EICAS) displays in the flight deck. During fueling operations, this information is also shown on the fueling indicators on the fueling control panel.

The FQIS processor basically consists of circuit cards, a volumetric top off(VTO) override switch and connectors. The FQIS processor contains seven fuel quantity circuit cards which monitor the fuel tank components and calculate the fuel quantity in each tank.

Individual tank circuitry is isolated by using only on card for each tank. Tank position

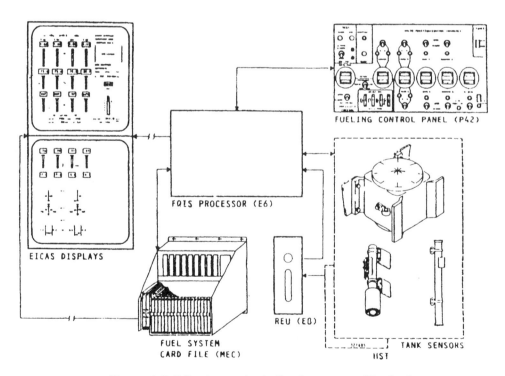

Figure 4-6-8 Fuel quantity indicating system(Boeing)

230 제4장 Aircraft System Description

is determined by pin locations in the processor card connectors. Several other circuit cards are used in the FQIS processor for interfacing with other fuel system components, control of the refuel valve relay and the volumetric top off adjustment cards.

Several of the fuel system electronic controls are circuit cards housed in the fuel system card file, shown in Figure 4-6-8. Each circuit card is mounted in one of the slots along the bottom shelf of the card file. Interfaces between the fuel system and these cards are made through connectors above the circuit card shelf. A system access panel provides direct access to some of these interface.

Each of the circuit card are line replaceable units(LRUs). Removal and replacement of these circuit card requires observance of standard electrostatic sensitive devices.

B. Flight Deck Fuel System Controls

The fuel management panel, shown in Figure 4-6-9, provides either direct or arming control of the crossfeed valves, boost pumps, and override/jettison pump. Light or

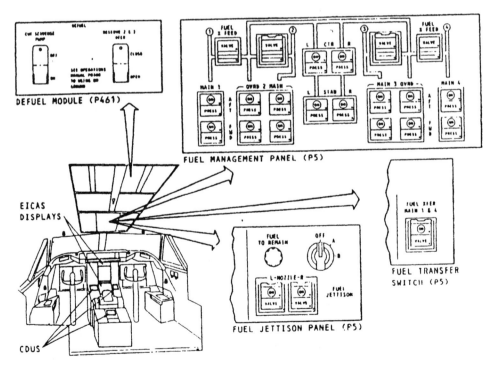

Figure 4-8-9 Fuel system control. Boeing 747-400 flight deck(Boeing)

4-6. Fuel Systems 231

mechanical bars in the panel switches provide display of system configuration. The panel is arranged as a schematic of the fuel system. Engines are noted number with fuel feed pathways shown interconnecting the tanks with engines.

Manual operation of a crossfeed valve is accomplished by pressing the respective fuel crossfeed valve switch. When pressed, the switch displays a bar which connects the fuel pathways on the panel schematic. When pressed again, the bar disappears. The amber crossfeed valve light in the switch comes on when the switch and valve positions disagree. crossfeed valve switches No.2 and 3 are guarded.

The other switches on the fuel management panel command operation of either boost pump or override/jettison pumps. Switch action for both types is the same. When the switch is pressed, and ON symbol appears on switch. When the switch is pressed again, the ON symbol disappears. Low discharge pressure is sensed at each pump outlet by a pressure switch. Low pressure indicator is shown by an amber light on the corresponding pump switch.

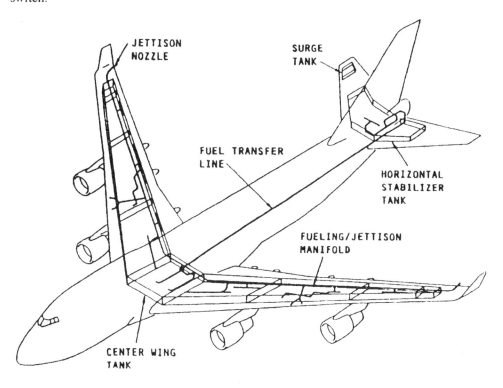

Figure 4-8-10 Horizontal stabilizer fuel transfer system (Boeing)

232   제4장 Aircraft System Description

The fuel transfer switch provides manual control of the fuel transfer valve between the outboard main fuel tanks and the inboard main fuel tanks. The fuel jettison panel provides controls for activation of the automated jettison sequency. An adjustable control is included for setting the amount of fuel to remain.

The refuel panel provides manual control of the center wing tank(CWT) scavenge pump and the reserve tank transfer valves for refueling. During normal operation, electronic control activate these components. The EICAS displays provide fuel system indications and messages. Access to the central maintenance computer system(CMCS) is provided through the control display units(CDU).

Fuel transfer from the 3,300 gallon horizontal stabilizer tank to the center wing tank is semi-automatic in operation. The basic transfer system is shown in Figure 4-6-10. The transfer pumps are armed by the flight crew during preflight from switches on the over head panel. The pumps are turned on with a signal from the fuel system management cards when fuel in center wing tank is reduced to 10,000 gallons. At that time, the isolation valves automatically open. Fuel is then pumped from the stabilizer to the center wing tank.

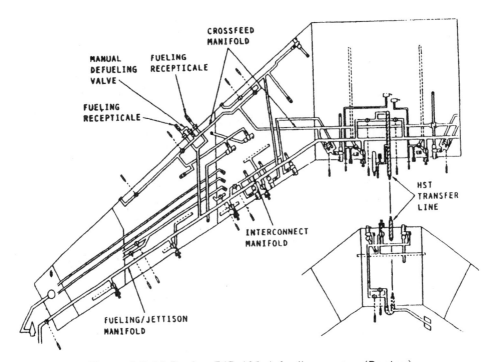

Figure 4-6-11 Boeing 747-400 defueling system(Boeing)

C. Defueling

The defueling system schematic, shown in Figure 4-6-11. shown the various paths the fuel can flow during selected operation, including defueling, and how the systems interface, allowing crossfeed and transfer. It should be noted that the fuel need not be pumped overboard to empty a specific tank, In many cases, fuel may be transferred into tank with available space.

Fuel in the reserve tank can only be transferred into the inboard main tank(2 & 3). This transfer is accomplished by gravity. The flow of fuel is controlled by two transfer valves installed in parallel. To defuel or transfer fuel from the main tanks, the boost pumps are used to pump the tanks down to sump level.

4-7. Hydraulic Systems

Although some aircraft manufacturers make greater use of hydraulic systems than others, the hydraulic system of the average modern large transport aircraft performs many functions. Among the units commonly operated by hydraulic systems are landing gear, leading edge devices, flaps, speedbrakes, wheel brakes, and flight control surfaces.

Hydraulic systems have many advantages as a power source for operating various aircraft units. Most hydraulic systems are similar, regardless of their function. Much of the information learned about a particular system can be generalized to study other systems.

1) Hydraulic Fluid

Hydraulic System liquids are used primarily to transmit and distribute forces to various units to be actuated. Liquids are able to do this because they are almost incompressible. Thus, if a number of passages exist in a system, pressure can be distributed through all of them by means of hydraulic fluid.

Manufacturers of hydraulic devices usually specify the type of hydraulic fluid to be used. Some of the properties and characteristics that must be considered when selecting a satisfactory liquid for a particular system are viscosity, chemical stability, fire and flash point.

A. Viscosity

One of the most important properties of any hydraulic fluid is its viscosity. Viscosity is internal resistance to flow. A liquid such as gasoline flows easily(has a low viscosity) while a liquid such as tar flows slowly(has a high viscosity). Viscosity increases with temperature decreases.

Chemical stability is another property which is exceedingly important in selecting a hydraulic liquid. It is the liquid's ability to resist oxidation and deterioration for long periods. All liquids tend to undergo unfavorable chemical changes under severe operating conditions. This is the case, for example, when a system operates for a considerable period of time at high temperatures.

Flash point is the temperature at which a liquid gives off vapor in sufficient quantity to ignite momentarily or flash when a flame is applied. A high flash point is desirable for hydraulic liquids because it indicates good resistance to combustion and a low degree of evaporation at normal temperatures.

Fire point is the temperature at which a substance gives off vapor in sufficient quantity to ignite and continue to burn when exposed to a spark or flame. Like flash point, a high fire point is required of desirable hydraulic liquids.

B. Phosphate Ester Base Fluid

Currently used in many large aircraft is Skydrol® 500B, a clear purple liquid having good low temperature operating characteristics and low corrosive side effects ; and Skydrol® LD, a clear purple low weight fluid formulated for use in large transport aircraft.

Test proved non-petroleum base fluids(Skydrol®)would not support combustion. Even though they might flash at exceedingly high temperatures. Skydrol® fluids could not spread a fire because burning was localized at the source of heat. Once the heat source was removed or the fluid flowed away from the source, no further flashing or buring occurred.

C. Intermixing of Fluids

Due to the difference in composition, hydraulic fluids will not mix. Neither are the seals for any one fluid useable with, or tolerant of, any of the other fluids. Should an aircraft hydraulic system be serviced with the wrong type fluid, immediately drain and flush the system and maintain the seals according to the manufacturer's specifications.

D. Health and Handling

Skydrol® fluid has a very low order of toxicity when applied to the skin in liquid form, but it will cause pain on contact with eye tissue. First aid treatment for eye contact includes the flushing the eyes immediately with large volumes of water and the application of any anesthetic solution. The individual should be referred to a physician.

In mist or fog form. Skydrol® is quite irritating to nasal or respiratory passages and generally produces coughing and sneezing, etc.

Silicone ointments, rubber gloves, and careful washing procedures should be utilized to avoid excessive repeated contact with Skydrol® in order to avoid solvent effect on skin.

2) Components of Hydraulic Systems

A. Basic Hydraulic System

Figure 4-7-1 shows a basic hydraulic system. The first of the components, the reservoir, stores the supply of hydraulic fluid for operation of the system. It replenishes the system fluid when needed and provides a means for bleeding air from the system.

A pump is necessary to create a flow of fluid. This basic system utilizes a hand pump to provide pressure. Large aircraft systems are, in most instances, equipped with engine-driven, electric motor-driven, engine bleed air-driven, or ram air turbine driven pumps. One system can also power another onboard system by using a power transfer unit.

The selector valve, shown in Figure 4-7-1, is used to direct the flow of fluid. These valves are actuated by solenoids or manually operated, either directly or indirectly through use of mechanical linkage. An actuating cylinder converts fluid pressure into useful work by linear or reciprocating mechanical motion, whereas a hydraulic motor converts fluid pressure into useful work by rotary mechanical motion.

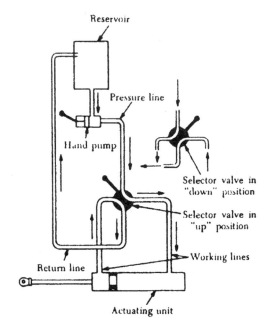

Figure 4-7-1 Basic hydraulic system with hand pump and four way selector valve

B. Filters

A filter is a screening or straining device used to clean the hydraulic fluid, thus preventing foreign particles and contaminating substances from remaining in the system. If such objectionable material is not removed, it may cause the hydraulic system to fail through the breakdown or malfunctioning of a single unit of the system.

There are many models and styles of filters. Their position in the aircraft and design requirements determine their shape and size. Most filters used are generally of the inline type. The inline filter assembly is comprised of three basic units: head assemble, bowl, and element. The head assembly is that part which is secured to the aircraft structure and connecting lines.

4-7. Hydraulic Systems

Within the head there is a bypass valve which routes the hydraulic fluid directly from the inlet to the outlet port if the filter element becomes clogged with foreign matter. The bowl is the housing which holds the element to the filter head and is that part which is removed when element removal is required.

The element may be either a micronic, porous metal, or magnetic type. A typical micronic type filter is shown in Figure 4-7-2. This filter utilizes an element made of specially treated paper which is formed in vertical convolutions(Folds). An internal spring holds the elements in shape.

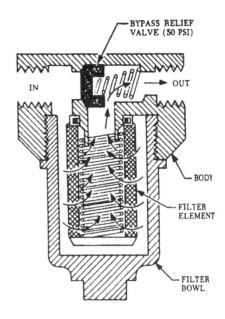

Figure 4-7-2 Hydraulic filter, micronic type

C. Power-Driven Pumps

Many of the hydraulic pumps used in aircraft are of the variable-delivery, or constant-delivery, type. A constant-delivery pump, regardless of pump RPM, forces a fixed, or unvarying, quantity of fluid through the outlet port during each revolution of the pump. Constant-delivery pumps are sometimes called constant-volume or fixed delivery pumps.

A variable-delivery pump has fluid output that is varied to meet the pressure demands of the system by varying its fluid output. The pump output is changed automatically by a pump compensator within the pump.

Various types of pumping mechanisms are used in hydraulic pumps, such as gears, gerotors, vanes, and pistons. The piston-type mechanism is commonly used in power-driven pumps because of its durability and capability to develop high pressure. In 3,000 PSI hydraulic systems, piston-type pumps are widely used.

The basic pumping mechanism of piston-type pump(see Figure 4-7-3) consists of a multiple-bore cylinder block, a piston for each bore, and a valving arrangement for each bore. The purpose of the valving arrangement is to let fluid into and out of the bores as the pump operates.

238 제4장 Aircraft System Description

Hydraulic pumps, as mentioned earlier, can be driven by the engines, electric motors, air turbine motors, ram air turbines, or a power transfer unit. Engine-driven pumps are mechanically linked and mounted on the engine accessory case.

Most electric motor driven pumps use alternating current to power the electric motor which drives the hydraulic pump. Switches in the cockpit control power to the electric motor-driven pumps.

Another method of powering hydraulic pumps can originate as pneumatic power(APU, ground connection, engine bleed air) which is used to drive an air turbine that turns the hydraulic pump.

In case all other methods of powering the hydraulic systems are lost, a ram air turbine(RAT) can be pump. Ram air turbines are generally used only in emergency situations. The RAT, shown in Figure 4-7-4, drops down and the turbine(propeller) begins to windmill which turns a hydraulic pump to pressurize one of the hydraulic systems.

Power transfer units are basically two hydraulic pumps mounted back to back with one pump functioning as a motor. For example, with one system pressurized, half the unit acts as a motor, and the other half of the unit acts as a pump. Power transfer units(PTUs) are used to pressurize one system to another. The transfer of power is mechanical, as no fluid passes between systems. PTU's can be used to power hydraulic systems during ground operations, or in flight if needed.

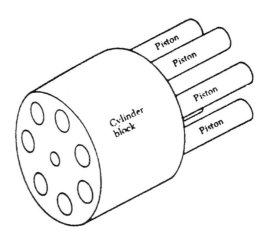

Figure 4-7-3 Axial-piston pump

D. Pressure Relief Valves

A pressure relief valve is used to limit the amount of pressure being exerted on a confined liquid. This is necessary to prevent failure of components or rupture of hydraulic lines under excessive pressures. The purpose of the system pressure gauge is to measure the pressure, in the hydraulic system, used to operate hydraulic units on the aircraft.

4-7. Hydraulic Systems   239

E. Accumulator

The accumulator is a steel sphere divided into two chambers by a synthetic rubber diaphragm. The upper chamber contains fluid at system pressure, while the lower chamber is charged with air. The main functions of an accumulator are to dampen pressure surges, to supplement the power pump, and to store power for limited operation of hydraulic unit when the pump is not operating.

F. Actuating Cylinders

An actuating cylinder transforms energy in the form of fluid pressure into mechanical force, or action, to perform work. It is used to impart powered linear motion to some movable object or mechanism. Actuating cylinders are generally of the single-action or double-action type. In a single-action actuating cylinder, fluid under pressure enters the inlet port and pushes against the face of the piston, forcing the piston to move. As the pistion moves, air is forced out of the spring chamber through the vent hole, compressing the Spirng.

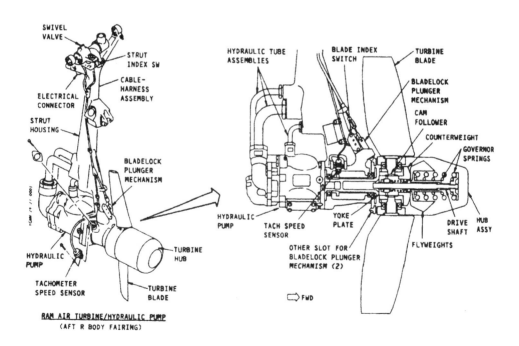

Figure 4-7-4 Ram air turbine/hydraulic pump(Boeing)

A double-action(Two-port) actuating cylinder is illustrated in **Figure 4-7-5.** The operation of a doule action actuating cylinder is ususally controlled by a four way selector valve. By allowing pressurized fluid to enter one of the ports, the piston and rod will be moved in the direction of the force being applied against the piston. In this case, the piston can be moved under pressure right or left, depending upon which fluid port is pressurized.

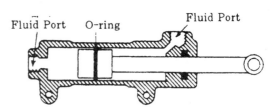

Figure 4-7-5 Double-action actuating cylinder

3) Boeing 757 Hydraulic System

Three independent, full-time, 3,000 PSI systems, using a synthetic BMS 3-11 type Ⅳ fluid, provide power for operation of landing gear, flight control and thrust reverser systems.

Each system is normally powered by two hydraulic pumps which are driven from independent power sources, Distribution of pressure from the three systems is such that the failure of one system will not result in loss of any flight control functions and the airplane can be safely operated in the event of loss of two hydraulic systems. An emergency hydraulic pump provides flight control operation in event of dual engine failure.

An electric motor pump is available in each system for ground maintenance, A central fill point facilitates fluid servicing of all three systems. Reservoir pressurization is obtained from the airplane pneumatic system and is available whenever the pneumatic ducts are pressurized. External hydraulic power can be connected to each system.

The three systems are color-coded to facilitate identification of tubing and components. Left system-red, center system-blue, and right system-green.

A. Hydraulic Power Distribution

The three hydraulic systems, take their names from their locations: left, right and center. The systems are powered by a total of 7 pumps. Multiple pumps in each system insure reliability, as illustrated in Figure 4-7-6.

The left and right are similar, with each system containing one engine driven pump(EDP) and one alternating current motor pump(ACMP). A power transfer unit(PTU)

4-7. Hydraulic Systems 241

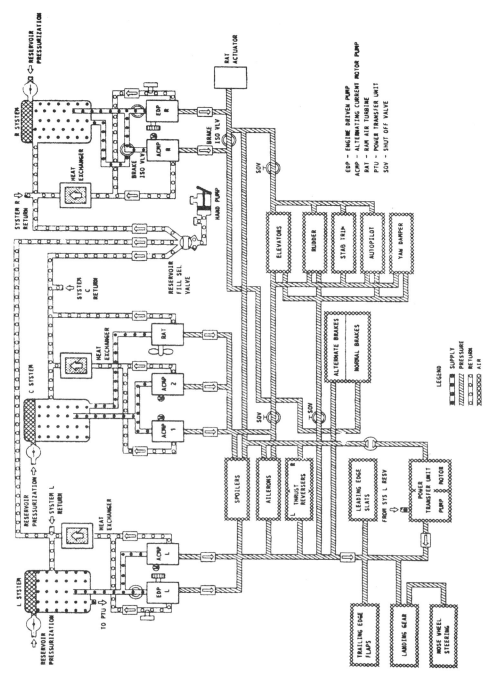

Figure 4-7-6 757 hydraulic system schematic(Boeing)

242 제4장 Aircraft System Description

connects the L & R systems mechanically. A hydraulic motor in the right system powers a hydraulic pump in the left system to provide sufficient flow to retract the landing gear and lift devices in the event of loss of the left engine or left engine driven pump. The ram air turbine(RAT) retract actuator is powered by the right system.

The center system has two ACMPs for primary pumps, and a ram air turbine(RAT) for emergency power. The components of the system are located in the wheel wells and body fairings. There is no hydraulic interconnection between the three systems.

The hydraulic indicator and control panel is located on the left side of the overhead panel. Controls and indicators include amber system low pressure lights, reservoir low quantity/pressure lights, pump low pressure lights, pump control switches and pump overheat lights as shown in Figure 4-7-7.

The ram air turbine(RAT) control switch, containing green pressure and amber unlocked lights, is located on the engine START/RAT control panel in the center of the overhead panel.

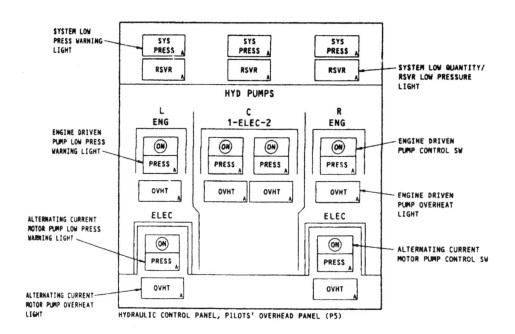

Figure 4-7-7 757 hydraulic control panel(Boeing)

B. System Operation

Each reservoir is pressurized from the pneumatic system by regulated air which is ported to the reservoirs. This air pressure is used to provide a positive head of pressure at the suction inlet of all the pumps.

A dual pressure switch on the reservoir provides an engine indicating and crew alerting system(EICAS) advisory annunciation for low presure indications.

Head exchangers in each system cool fluid returning to the reservoir. A fin-type heat exchanger, shown in Fugure 7-8, is installed in the main fuel tanks. The hydraulic flow from the case drain of the pumps within each system is cooled in the heat exchanger by thermal transfer between the hot hydraulic fluid and the cool fuel. The cooled hydraulic fluid is then returned to the system's reservoir. The heat exchangers must be completely immersed in fuel to provide adequate cooling.

Since the left and right hydraulic systems are very similar, only the right system's operation will be described. Both pumps, in the right system, shown in Figure 4-7-9, receive

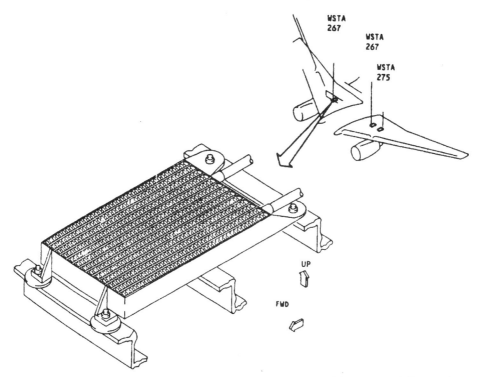

Figure 4-7-8 Hydraulic fluid heat exchanger mounted in fuel tank(Boeing)

244 제4장 Aircraft System Description

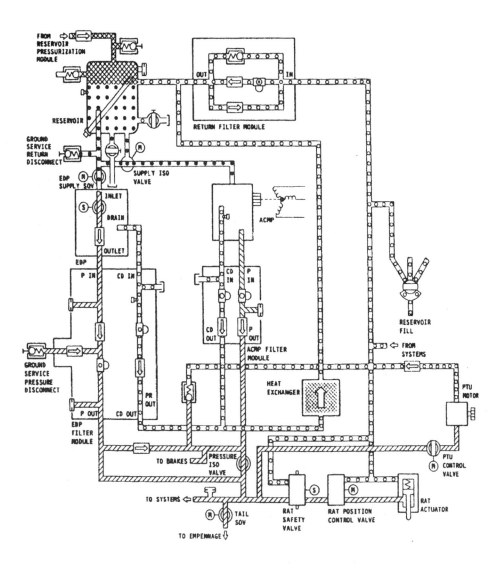

Figure 4-7-9 757 Right hydraulic system flow schematic(Boeing)

fluid from the reservoir standpipe and the ACMP can also receive fluid from the bottom of the reservoir when the reservoir isolation valve is operated. Both pumps are serviced by filter modules containing filters, pressure switches and check valves(see Figure 7-10).

4-7. Hydraulic Systems 245

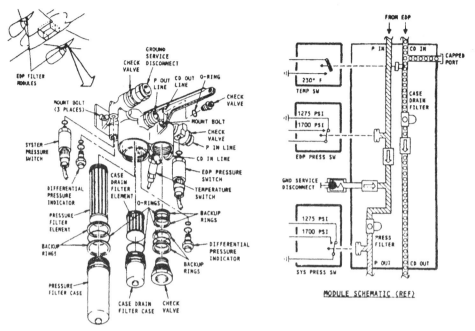

Figure 4-7-10 757 Engine driven pump filter module(Boieng)

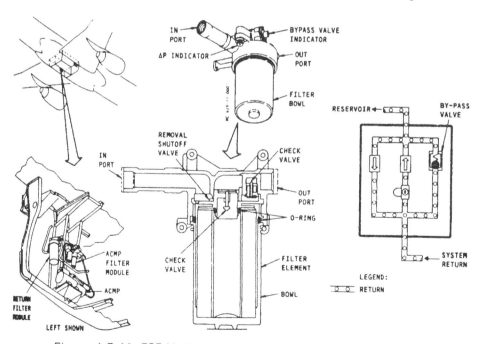

Figure 4-7-11 757 Hydraulic system return filter module(Boeing)

246 제4장 Aircraft System Description

The case drain(CD) flow from the pumnps passes through the heat exchanger mentioned earlier and returns to the reservoir. Return fluid from the systems is routed through a return filter. The return filter, shown in Figure 4-7-11, incorporates a removable shutoff valve to facilitate replacement of the 15 micron paper filter, and a bypass valve to permit bypassing a clogged filter.

A system relief valve prevents overpressure from the EDP and ACMP, which also powers the power transfer unit(PTU) motor. The power transfer unit(PTU) powers the left hydraulic system landing gear and flap/slat systems, utilizing the right hydraulic system engine driven pump(EDP), to provide operation of these systems when the left hydraulic system EDP is off or inoperative.

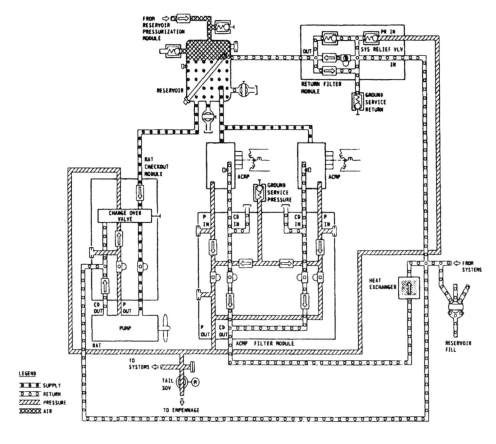

Figure 4-7-12 Center hydraulic system flow schematic, 757(Boeing)

4-7. Hydraulic Systems   247

A pressure isolation valve operates to dedicate ACMP pressure output to the brakes, if needed. Each system contains pressure and return ground service connections.

The center hydraulic system, shown in figure 7-12, uses a pressurized reservoir to provide fluid to two alternating current motor pumps and a ram air turbine pump. The ACMPs receive fluid from the reservoir standpipe and the RAT pump receives fluid from the bottom of the reservoir. The ACMPs and the RAT are serviced by modules containing filters. The fluid from the pumps passes through a heat exchanger and joins system return fluid flowing back to the reservoir. Any one of the pumps can pressurize the center hydraulic system.

4) Boeing 757 Landing Gear system

A. Main Gear

The main gear and nose gear incorporate a standard air-oil strut for shock absorption and for supporting the airplane's weight. Each gear is hydraulically extended and retracted and incorporates a hydraulically operate main gear door. The main gear is hydraulically tilted aft to fit into the wheel well structure on gear retraction and to provide air/ground sensing.

The main gear is held up and locked by an uplock hook engaging a roller on the shock strut, and held down and locked by overcenter locking of a downlock link. The main gear door actuator locks the main gear door closed.

Each gear has four wheels and brakes on a dual axle truck. The bearing mounted brakes are hydraulically actuated with antiskid protection provided.

Alternate extension(emergency) is accomplished by an electric/hydraulic system which unlocks the main gear and doors to allow free fall extension. Gear position indication is provided by a dual proximity switch system controlled by the proximity switch electronics unit.

B. Nose Gear

The nose gear is hydraulically retracted, free fails to extend, and incorporates hydraulically sequenced forward doors for aerodynamic seal. Overcenter locking of a lock link, hydraulically actuated and aided by a pair of bungee springs, locks the gear in both the extended and the retracted positions.

Hydraulically powered nose wheel steering for ground directional control is porvided

248 제4장 Aircraft System Description

with tiller or rudder control. Friction pads brake the nose wheels on retraction. Alternate extension and gear position indication is accomplished in much the same manner as the main gear.

C. Landing Gear Controls and Indicators.

A three position(UP,OFF,DN) landing gear lever, is used to control landing gear extension and retraction as shown is Figure 4-7-13. A lock solenoid in the landing gear lever prevents moving the lever to the up position when the main gear trucks are not tilted, A lock override button is provided. A guarded alternate EXTEND switch controls an electric motor-hydraulic pump which will unlock the gear doors and gear to allow free fall extension. Position indicators above the landing gear lever include three green gear down and locked lights, a gear door open amber light, and a gear disagreement amber light.

Eight hydraulic, brake assemblies are operated by either the captain's or first officer's brake pedals. The autobrake system is controlled by a rotary selector switch, with an amber light above the switch to indicate a disarm condition. Brake pressure is indicated by a

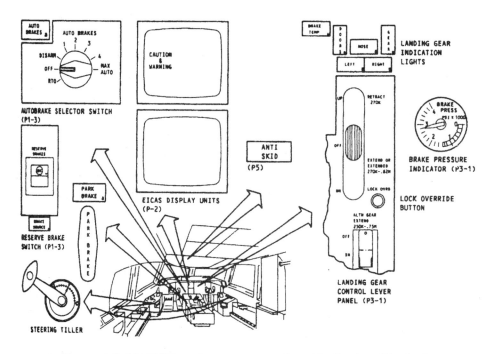

Figure 4-7-13 757 Landing gear control and indications(Boeing)

4-7. Hydraulic Systems 249

gauge handle on the quadrant stand. Indication of parking brake operation is provided by an amber light forward of the handle.

A reserve brakes switch will isolate the right hydraulic system alternating current motor pump to the brakes. An amber BRAKE SOURCE light, below the reserve brake switch, provides indication of loss of normal and alternate hydraulic brake sources. Thermocouple device on the brakes provide brake temperature sensing for display on EICAS. Another amber light indicates antiskid faults, with all amber lights having associated EICAS messages.

D. Nose Gear Steering

The boeing 757 nosewheel steering system, shown in figure 7-14, is designed to achieve airplane directional control for ground operations. The nosewheel steering is controlled by two steering tillers, located on the left and right side of the flight compartment, or by the rudder pedals. Using the tillers provides for turns up to 65 degrees left or right of center. The rudder pedals will give 7 degrees left or right.

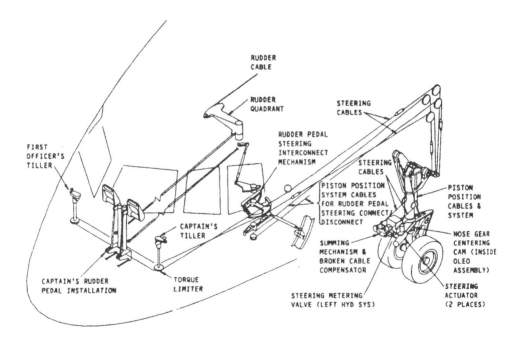

Figure 4-7-14 Nose wheel steering system description(Boeing)

250 제4장 Aircraft System Description

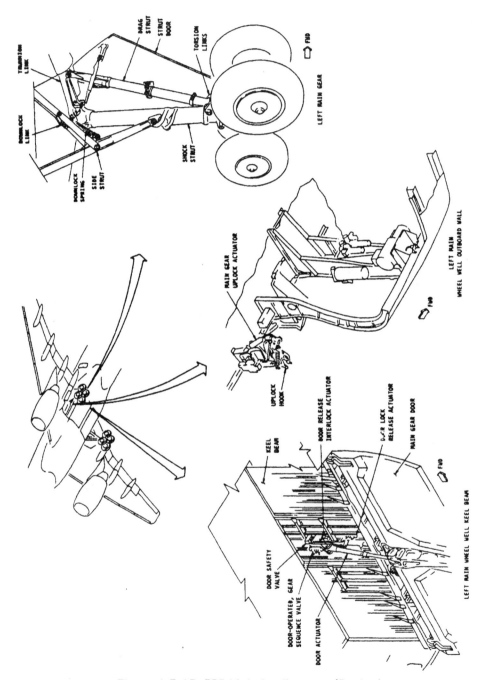

Figure 4-7-15 757 Main landing gear(Boeing)

4-7. Hydraulic Systems 251

whether the steering command is from the tillers or rudder system, the command signal is transmitted by cables to a hydraulic metering valve located on the nose gear. The metering valve directs hydraulic pressure to two steering actuators to steer the nose gear wheels. Internal centering cams in the nose gear shock strut center the wheels when the strut is extened after takeoff and unpressurized during flight.

E. Main Landing Gear Structure

The main landing gear structure consists of a shock strut, torsion links, truck assembly, trunnion link, drag strut, side strut and down lock assembly, as can be seen in Fighre 7-15. The shock strut inner and outer cylinders perform standard air-oil shock absorption and are serviced with dry air or nitrogen through a gas chaging valve on the top of the strut, and with oil through an oil charging valve on the aft side of the strut. Torsion links connect the shock strut inner and outer cylinders. The truck assembly consists of truck beam, axles, brake rods, and a protective shields. The truck beam attaches to the bottom of the inner cylinder, providing the pivot and attach point for the truck assembly.

F. Nose Landing Gear Structure

The nose gear structure, shown in Figure 4-7-16, consists of a shock strut, torsion links, drag brace, and lock links. The shock strut is trunnion mounted in the nose gear wheel well and is supported by a two piece folding drag brace.

The upper drag brace is trunnion mounted to the wheel well structure and the lower brace attaches to a forging on the shock strut outer cylinder. The drag brace is held locked in both the extended and retracted positions by overcenter locking of the drag brace and the aft link to a fitting on the aft nose wheel well bulkhead.

Bungee springs and a hydraulic actuator provide sequencing and overcenter locking of the lock links, which are responsible for locking the gear in the extended and retracted positions.

G. Proximity Switch System

The system monitors the position of the landing gear and other aircraft components. The proximity switch electronic unit(PSEU) contains the microprocessor and circuit cards to provide position indication and fault annuncition for the monitored systems by utilizing the feedback from proximity sensors.

The proximity switch electronics unit(PSEU) is a digital control unit containing four

252 제4장 Aircraft System Description

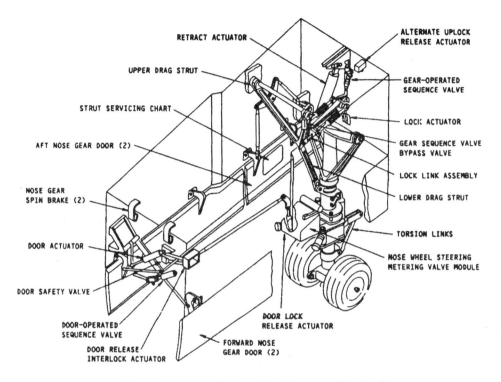

Figure 4-7-16 757 Nose landing gear(Boeing)

different types of circuit cards; proximity, logic driver, and BITE cards. There are six interchangeable proximity cards. Each card contains 16 channels which pulse the sensors, determine target near or far, and switch voltage to the logic cards to a high or low output.

The logic cards receive the high and low voltage signals from the proximity cards and provide outputs to the driver cards which provide grounds(completed circuits) to lights. relays, switches, EICAS, etc. Two microprocessor based, built-in test equipment(BITE) cards provide fault isolation and system test capability.

Sensors provide position information to the proximity switch electronics unit(PSEU). Each sensor is a two-wire magnetic field producing coil-core device with wires connected to the PSEU proximity switch circuit card. The sensor operates in conjunction with a steel(magnetic) target. This action can be seen in Figure 4-7-17. The change in inductance as the proximity of sensor and target vary, provides a high or low output signal from the proximity switch cards to the appropriate logic card.

4-7. Hydraulic Systems   253

H. Air/Ground System

Air/ground realys switch various airplane system control circuits from ground to air mode, and air to ground mode. The relays are controlled by the proximity switch electronic unit(PSEU) using inputs from the main gear truck tilt proximity sensors, the nose gear compressed proximity sensors, and truck positioner shuttle valve pressure switches.

Tow sensors on each main gear truck provide dual system truck tilt inputs to the PSEU. Two sensors on the nose gear strut provide dual nose gear strut compression inputs to the PSEU.

The sensor inputs are processed in the PSEU logic to provide inputs to drive air/ground realys which control various air/ground critical systems. PSEU and air/ground relay outputs are provided to the EICAS computers for air/ground system fault detection and annunciation.

I. Gear Position and Indication

Position indication for the landing gear system is provided by green and amber lights and EICAS caution, advisory, status and maintenance messages. The proximity switch

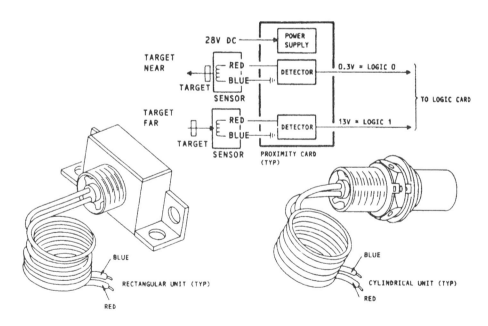

Figure 4-7-17 Proximity system sensors and targets(Boeing)

electronic unit(PSEU) controls the lights and messages, using inputs from dual proximity sensor systems(System 1 and System 2).

An amber GEAR light illuminates when the landing gear position is not in agreement with the landing gear lever position, as sensed by both system 1 and system 2. Three green lights, NOSE, LEFT and RIGHT, illuminate when the respective landing gear is down and locked. An amber GEAR DOORS light illuminates when a landing gear door is not closed as sensed by both system 1 and system 2.

J. Wheels and Brake Assemblies

The main landing gear consists of dual tandem trucks, four wheels per truck. Each of the eight main gear wheels is provided with a hydraulically actuated, multiple disc brake. The nose landing gear is a single strut with dual wheels. The nose gear wheels do not incorporate brakes

The normal brake system, shown in Figure 4-7-18, is powered by the right hydraulic system. The alternate system is powered by the left hydraulic system and is automatically selected upon loss of the right hydraulic system pressure. An accumulator in the right(normal) system is automatically selected when both normal and alternate system pressure is lost.

A reserve braking system is included in the normal brake system. The normal brake metering valves control brake pressure to the autobrake shuttle valves and the normal antiskid valves. The alternate brake metering valves meter hydraulic system pressure directly to the alternate antiskid valves.

Landing gear up line pressure is ported to actuators on the alternate antiskid valves to stop wheel rotation during landing gear retraction. Hydraulic pressure from the antiskid valves passes through a shuttle valve module which selects the pressurized system to the wheel brake assemblies.

When the right hydraulic system is pressurized, the accumulator is charged and the accumulator isolation valve(AIV) is open. A control line directs pressure to close the alternate brake selector valve(ABSV). Brake pedal operation meters right system pressure through the autobrake shuttle valves to the normal antiskid modules. The normal antiskid modules control hydraulic pressure individually to each brake through a shuttle valve module.

4-7. Hydraulic Systems 255

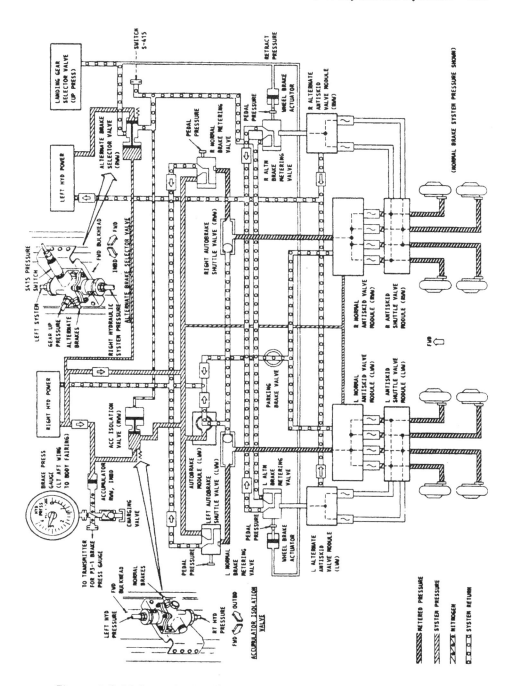

Figure 4-7-18 Brake hydraulic system, normal condition, 757(Boeing)

256 제4장 Aircraft System Description

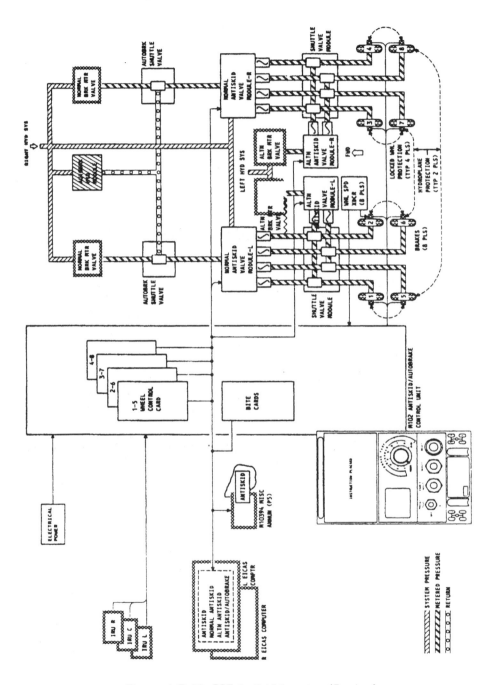

Figure 4-7-19 757 Antiskid system(Boeing)

4-7. Hydraulic Systems

K. Brake Temperature Monitoring System

The purpose of the monitoring system is to provide an indication of brake overheat by providing a color-coded number and box on the EICAS status page which represents the temperature of each brake and a white brake temperature light to indicate brake overheat.

L. Antiskid/Autobrake

The purpose of the antiskid system is to monitor wheel deceleration and provide brake release to achieve optimum braking action under varying braking conditions. Normal and alternate antiskid valves receive pilot metered brake pressure from normal or alternate brake metering valves, depending on which brake system is pressurized. The antiskid valves are controlled by the antiskid/autobrake control unit to provide brake release if excessive wheel deceleration is detected. This brake release signal is developed by the control unit from the individual wheel speed transducer inputs.

The control unit also provides touchdown and hydroplane protection using inertial reference system(IRS) ground speed inputs.

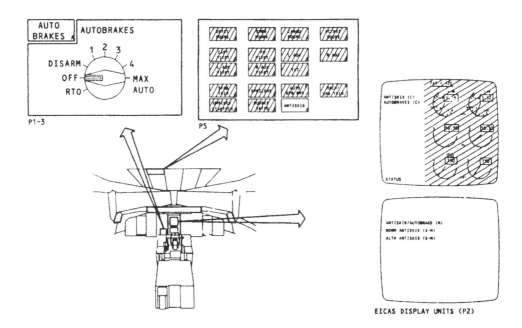

figure 4-7-20 Autobrakes flight deck controls and indicators(Boeing)

제4장 Aircraft System Description

There is no control switch for the antiskid system, therefore it is operational whenever the electrical systems are powered. An amber ANTISKID light indicates a fault in the antiskid system.

The antiskid function is controlled by four wheel control cards within the antiskid/autobrake control unit. Each card controls a fore/aft pair of wheels through individual wheel circuits, as can be seen in Figure4 7-19.

The antiskid valve modules receive metered pressure from the brake metering valves or autobrake module. If a skid condition is detected, the control unit will reduce brake pressure to that wheel by providing signals to the normal and alternate antiskid valves.

The autobrake system automatically applies and controls brake pressure in an attempt to achieve an airplane rate of deceleration, as selected by the flight crew. The antiskid/autobrake control unit operates an autobrake module to provide metered pressure to the brakes through the normal antiskid valves.

Brake pressure varies according to the rate of airplane deceleration selected and the actual rate of deceleration obtained through braking, thrust reverser and ground speedbrake operation.

The autobrakes are controlled through an autobrake selector, shown in Figure 4-7-20. The switch allows deceleration selections for landing, autobrakes, one selection for rejected takeoff(RTO), and OFF position and a disarm fault position. The amber AUTOBRAKE light indicates the system has faulted to a disarm mode.

4-8 Oxygen System

Without oxygen, people lose consciousness and can die in a short period of time. But, before this extreme state is reached, reduction in normal oxygen supplies to the tissues of the body can produce important change in body functions and processes.

The sluggish condition of mind and body caused by a deficiency or lack of oxygen is called hypoxia. There are several causes of hypoxia, but the one which most concerns aircraft operations is the decrease in partial pressure of the oxygen in the lungs which occurs at high altitudes.

The rate at which the lungs absorb oxygen depends upon the oxygen pressure. The pressure that oxygen exerts is about one-fifth of the total air pressure at any one given level. At sea level, this pressure value is sufficient to saturate the blood. However, if the oxygen pressure is reduced, either from the reduced atmospheric pressure at altitude, or because the percentage of oxygen in the air breathed decrease, then the quantity of oxygen in the blood leaving the lugs drops, and hypoxia follows.

At high altitude there is decreased barometric pressure, resulting in decreased oxygen content of the inhaled air, consequently, the oxygen content of the blood is reduced.

The low partial pressure of oxygen, low ambient air pressure, and temperature at high altitude make it necessary to greate the proper environment for passenger and crew comfort. The most difficult problem is maintaining the correct partial pressure of oxygen in the inhaled air. This environment can be achieved by using a pressurized cabin.

Pressurization of the aircraft cabin is now the accepted method of protecting persons against the effects of hypoxia. Within a pressurized cabin, people can be transported comfortably and safely for long periods of time, particularly if the cabin altitude is maintained at 8,000 feet or below, where the use of oxygen equipment is not required. However, the flight crew in transport aircraft must be aware of the danger of accidental loss of cabin pressure(explosive or rapid decompression) and must be prepared to meet such an emergency, should it occur.

Transport aircraft are equipped with supplemental, or emergency, oxygen systems in case of cabin decompression for both the flight crew and the passengers. Emergency supplemented oxygen is a necessity in any pressurized aircraft flying above 25,000 feet. The flight crew and passenger oxygen systems are generally separate and independent of each other in their operation.

The flight crew oxygen system is supplied from cylinders of compressed oxygen that

260 제4장 Aircraft System Description

flows through shutoff valves, regulators, lines, and masks to the flight crew. Under certain operating situations and decompression, the flight crew uses the oxygen mask for breathing. The flight crew oxygen system is usually a pressure-demand(on-demand) system.

The system provides either pure 100%, oxygen or diluted oxygen(a mixture of oxygen and air from the flight deck environment), to the flight crew's oxygen mask. When one of the flight crew members inhales, oxygen is provided, but only when a breath is drawn in. The amount the oxygen is diluted with flight deck air will depend upon the flight deck altitude. This type of system is sometimes referred to as a diluter-demand system. Flight crew oxygen systems will be discussed in more detail later in this chapter.

Passenger emergency oxygen systems ar generally of the continuous flow type which provides a steady flow of oxygen to the mask. Passenger oxygen systems have oxygen mask that drop from the ceiling of the cabin in case of cabin decompression. There are generally two types of systems, the compressed oxygen cylinder, or the solid state oxygen system(chemical oxygen generator).

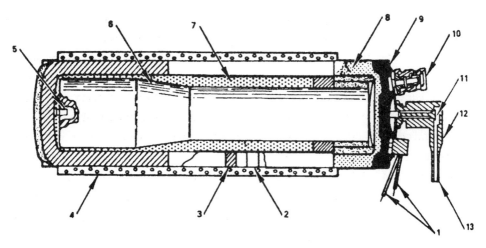

1. ELECTRICAL LEADS
2. WHITE STRIPE
3. BLACK COMPARISON STRIPE
4. SHIELD
5. IGNITER ASSEMBLY
6. CHEMICAL CORE
7. CORE SUPPORT
*8. HOPCALITE
9. FILTER PAD
10. RELIEF VALVE ASSEMBLY (50 TO 75 PSI)
11. OUTLET PORT
12. MANIFOLD ASSEMBLY
13. MANIFOLD ASSEMBLY OUTLET

*A GRANULAR MIXTURE OF THE OXIDES OF COPPER, COBALT, MANGANESE AND SILVER USED IN GAS MASKS TO CONVERT CARBON MONOXIDE TO DIOXIDE.

Figure 4-8-1 Chemical oxygen generator-cutaway view(Lockheed)

4-8. Oxygen Systems   261

The chemical oxygen generator differs from compressed oxygen cylinder, in that the oxygen is actually produced at time of delivery. Sodium chlorate, when heated to 478 degrees F. releases up to 45% of its weight as gaseous oxygen. The necessary heat for decomposition of the sodium chlorate is supplied by iron which is mixed with the chlorate. The sodium chlorate and iron power make up oxygen generator chemical core. The initiator assembly starts the burning of the chemical core, and the oxygen produced is filtered before leaving the oxygen generator through the outlet port. A cutaway view of an oxygen generator is shown in Figure 4-8-1.

Oxygen furnished at the mask is odorless, tasteless, and has a temperature of approximately 80 degrees F. The oxygen generator case temperature reaches approximately 400 degrees F. The white stripe will darken to show that the oxygen generator has been used. Normal operating pressure is below 50 PSI before being reduced for the mask.

The chemical generator can be initiated, or ignited, by activating the squib(electrically as on the L-1011) or by a percussion device. When a percussion device is used, passengers pulling the mask down for use releases a firing pin which starts the generator. The generators provide oxygen flow for about 12 to 18 minutes.

1) Lockheed L-1011 Oxygen system

A. Flight Crew Oxygen system

The crew system, shown in figure 4-8-2, will last a 5-man crew approximately 4 hours using diluted oxygen. An overpressure relief fitting will release all cylinder contents overboard if cylinder pressure exceeds a certain limit. If this happen, the overboard discharge indicator, green plastic disc mounted at the skin line on the right hand side of the flight station, will blow out.

If the oxygen cylinder's temperature is not excessive, an overpressure of the oxygen cylinder would normally be very rare. A slow-opening on-off valve releases cylinder pressure to the pressure reducer. The pressure reducer decreases cylinder pressure(1850 PSI at 70 degrees F.) to 50-90 PSI. It also contains a relief valve that will safely relieve momentary overpressure.

On a panel near the cylinder is a quick disconnect fitting for maintenance checks. Corrosion-resistant steel tubing distributes the reduced pressure oxygen to five panel-mounted diluter-demand regulators, which further reduce oxygen pressure to breathing pressure level.

262 제4장 Aircraft System Description

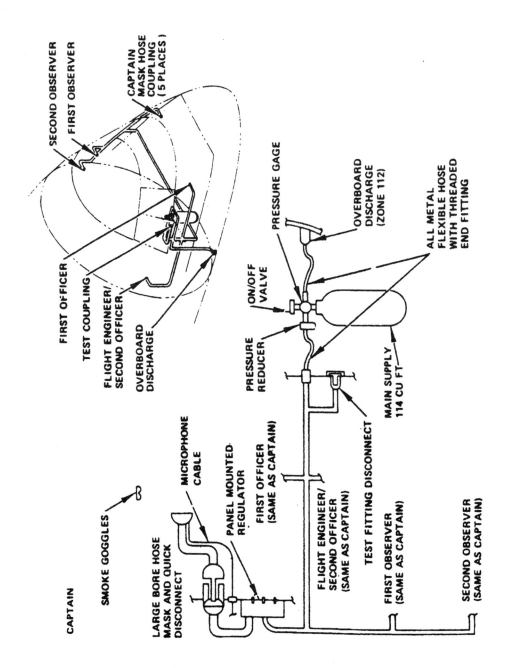

Figure 4-8-2 L-1011 Crew oxygen system schematic diagram(Lockheed)

4-8. Oxygen Systems

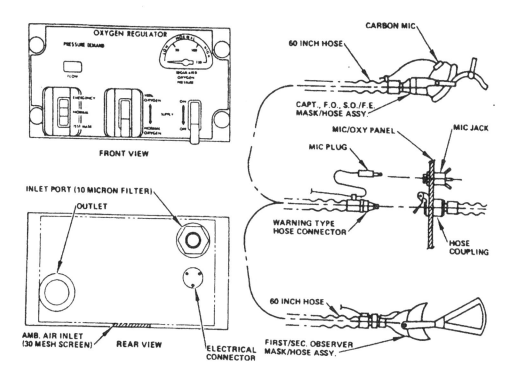

Figure 4-8-3 Flight crew oxygen mask, hose, and diluter-demand regulator(Lockheed)

These regulators indicate distribution line pressure, and contain a flow indicator and toggle control valves for oxygen pressure and dilution control Aneroid within the regulators lessen oxygen dilution with increasing altitude and provide undiluted oxygen at positive pressure above 28,000 feet flight station altitude.

Near the panel-mounted regulators are mask hose quick disconnects(QDs) and microphone cable jacks. These hose QDs will also accept the hose with a full-face smoke mask supplied with the flight station portable oxygen cylinder. To determine the oxygen level in the cylinder, it is weighted.

The flight crew oxygen mask, hose, and diluter-demand regulator are shown in Figure 4-8-3. A panel-mounted oxygen regulator and a mic/oxy panel are located near each of five flight station crew seats. On the front of the oxygen regulator are controls for oxygen dilution and pressure. On the back of the oxygen regulator are oxygen inlet and outlet ports, and an electrical connector for the 5 VAC panel light. On the bottom of the oxygen regulator is an ambient air inlet. The end of the mask hose that connects to the mic/oxy panel has a warning type connector that will not allow oxygen to flow through it if the connector is not seated in the hose coupling.

The regulated oxygen pressure gauge shows pressure entering the oxygen regulator. Oxygen will not flow from the oxygen regulator to the masks unless the on-off selector is in the ON position shown on the right side of the oxygen regulator panel in Figure 4-8-3.

With the 100% oxygen-normal oxygen selector in the 100% position, undiluted oxygen will be supplied to the mask regardless of flight station altitude. With this selector in the normal oxygen position, oxygen to the mask will be diluted with flight station ambient air in proportion to flight station air pressure. Dilution decrease with increasing altitude. At a flight station altitude of 28,000 feet or higher, oxygen to the masks will not be diluted at all. Normal flight station altitude(pressurized) is 8,000 feet.

With the emergency-normal-test mask selector in the normal position, flow to the mask is on demand(inhalation). In the emergency position, flow to the mask is continuous, and at positive pressure. The emergency position mechanically overrides the normal position. In the test mask position, flow to the mask is continuous, and at a pressure higher than in emergency position. The selector is spring-loaded out of the test mask position to the normal position.

The flight crew is required to have a portable breathing oxygen cylinder equipped with a full face smoke mask that operates in much the same manner as the crew oxygen system. It also operates with an undiluted on-demand regulator as shown in Figure 4-8-4. The mask hose can be connected to any crew station quick disconneckt. Likewise, any crew mask can be connected to the portable cylinder. A pressure regulator at the top of the cylinder has a slow opening on-off valve, a charging valve, a pressure gauge, a relief valve an overpressure safety plug, and a capped constant flow outlet, which is not used.

B. Passenger Oxygen System

Emergency oxygen for passengers is supplied from chemical oxygen generators through conventional face-cup masks. The generators, which are initiated by an electrical pulse, burn the two chemicals and produce oxygen for 15 to 18 minutes. Generators and associated rack-mounted mask(s) are secured in modules, one about each passenger seat group, attendant seat, and in each lavatory ceiling.

Oxygen masks, with associated supply tubes, are nested in a clear plastic rack and secured in the module by a hinged door. The door is magnetically latched, and for mask

4-8. Oxygen Systems 265

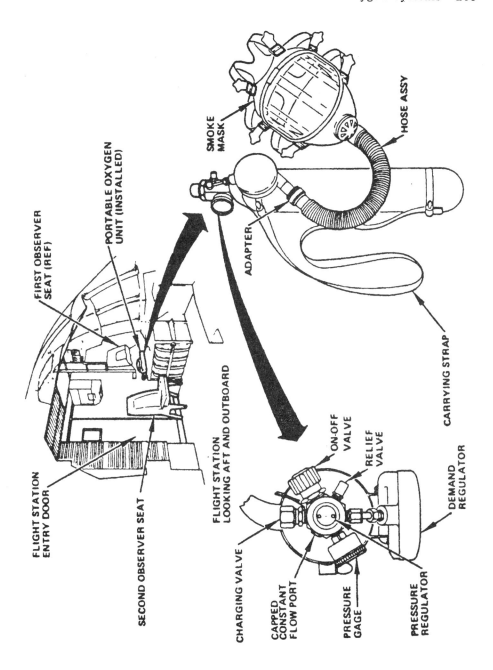

Figure 4-8-4 Flight crew portable breathing oxygen cylinder (Lockheed)

266 제4장 Aircraft System Description

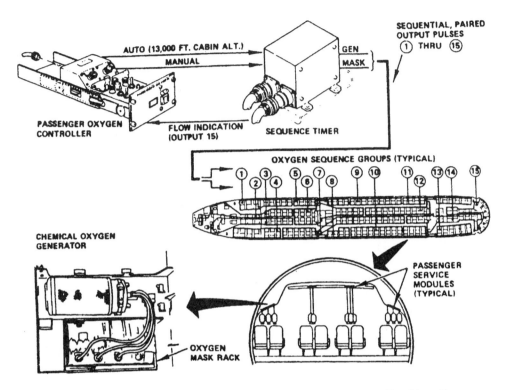

Figure 4-8-5 Lockheed L-1011 passenger oxygen system(Lockheed)

deployment, is electromagnetically unlatched by a solenoid. Oxygen generators and the masks associated with them, are divided into fifteen groups ; each guoup being identifyed numerically and composed of sectionally-sittuated modules.

System control components, shown in Figure 4-8-5, include a passenger oxygen controller, located at the Flight engineer's panel, and a sequence timer, located in the center cabin celling. The controller, through aneroid switching, automatically activates the system if cabin altitude rises to 13,000 feet. A manual switch on the controller provides from manual activation of the system if the aneroid switching circuit fail.

The sequince timer, when signalled automatically or manually, sequentially develops fifteen paired, timed, and synchronized output pules, each pair consisting of a generator initiation pulse and a mask drop pulse. The 15 paired out puts are completed in about 7 seconds, then are redundantly repeated. The fifteenth output of the first cycle illuminates the OXYGEN FLOW light on the controller. Aircraft wiring connects the 15 output pulse

4-8. Oxygen Systems   267

pairs to the modules in each sequence group, initiating each generator in the group and energizing the solenoid for unlatching the mask rack door. The passenger oxygen controller incorporates a test panel which contains provisions for testing the system.

C. Chemical Oxygen Generator

Oxygen generators are available in three sizes to supply oxygen to one, two, or three masks, as shown in Figure 4-8-6. They are approximately 7 inches long and 3 inches in diameter. Hermetically sealed cases have a shelf life of 10 years. Date of manufacture is indicated on the nameplate.

The case exterior contains black and white comparison stripes. The white stripe is heat sensitive and will darken when the generator is used.

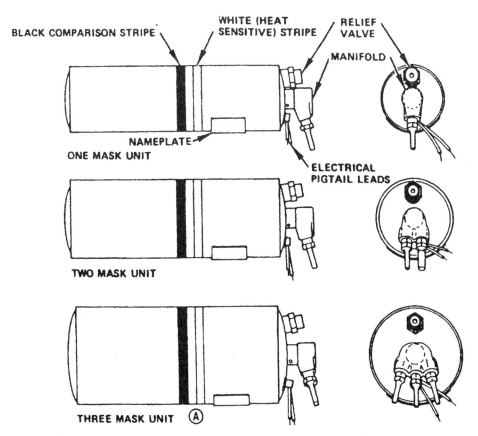

Figure 4-8-6 Chemical oxygen generator arrangement (Lockheed)

268 제4장 Aircraft System Description

Oxygen from all masks constitutes about 1 percent of the cabin atmosphere, so there is no fire hazard due to oxygen from unused masks. Oxygen flow is continuous until the generator is expended.

D. Passenger Service Modules

An example passenger service module(PSM), shown in Figure 4-8-7, is ceiling mounted above each set of two passenger seats on the left and right sides of the cabin centerline. Oxygen masks are rack mounted in a mask compartment in the passenger service module(PSM) and are retained by a latched door.

A chemically oxygen generator, mounted adjacent to the mask compartment, is manifolded by clear plastic tubes to each oxygen mask. When the passenger oxygen system is activated, the door solenoid is energized ; the door unlatches and springs open ; the masks drop and are suspended by their supply tubes. Simultaneously, the chemical oxygen generator is actuated and supplies oxygen to each mask through its supply tube.

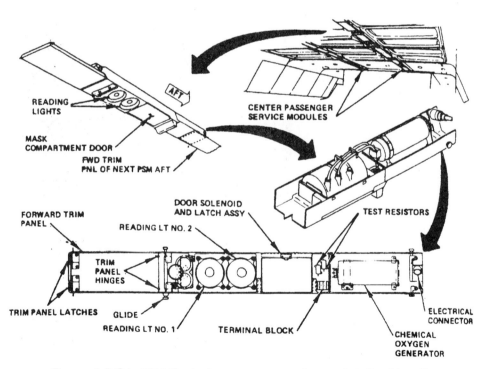

Figure 4-8-7 L-1011 Typical passenger service module(Lockheed)

4-9. Warning and Fire Protection Systems

Large transport aircraft have many on-board warning systems used to alert the flight crew to several unsafe operational conditions. Some of these systems include fire and overheat, unsafe takeoff, unsafe landing, overspeed, cabin pressure, altitude alert, and ground proximity warning systems. Other warning systems used to provide information to the pilots or the flight engineer can be doors, flight controls, slat position. auto-pilot disconnect and others.

Warning systems have generally the same basic function on most aircraft, although some warning systems do differ in configuration and operation. The particular warning systmes used can also vary from aircraft to aircraft.

On newer generation aircraft, most of the warning systems are fed into the Engine Indication and Crew Alerting System(EICAS), which provides messages or warnings to the flight crew on cathode-ray-tubes(TV screen) in the cockpit.

Along with fire detection warning systems, fire extinguishing systems also have a major role in protecting the aircraft against the threat of an onboard fire.

1) Fire Protection Systems

Because fire is one of the most dangerous threats to an aircraft, transport aircraft are protected by overheat, smoke, and fire protection systems. A complete fire protection system includes both a fire detection and a fire extinguishing system.

To detect fires or overheat conditions, detectors are placed in the various zones which are being monitored. A fire zone is an area or region of an aircraft which requires fire detection and/or fire extinguishing equipment and a high degree of inherent fire resistance. Some fire zone areas are the engine nacelles, APU areas, cargo compartments, and main landing gear wheel wells.

A fire detection system should signal the presence of a fire or overheating. Units of the system are installed in locations where there are greater possibilities of a fire, such as fire zones. Detector systems in common use are the thermal switch system and the continuous-loop detector system.

Thermal switches are heat sensitive units that complete electrical circuits at a certain temperature. They are connected parallel with each other, but in series with the indicator lights. If the temperature rises above a set value in any one section of the circuit, the thermal switch will close, completing the light circuit to indicate the presence of a fire or overheat condition. The thermal switch system uses a bimetallic thermostat switch or spot detector, similar to that shown in Figure 4-9-1. Each detector unit consists of a bimetallic thermoswitch. Most spot detectors are dual terminal thermoswitches.

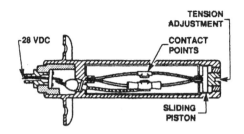

Figure 4-9-1 Fenwal spot detector

A continuous-loop detector or sensing system permits more complete coverage of a fire hazard area than any type of spot-type temperature detectors. Continuous-loop systems are versions of the thermal switch system. They are overheat systems;heat sensitive units that complete electrical circuits at a certain temperature. Two widely used types of continuous-loop systems are the Kidde and the fenwal systems.

In the Kidde continuous-loop system(Figure 4-9-2), two wires are imbedded in a special ceramic core within an inconel tube. One of the two wires in the Kidde sensing system is welded to the case at each end and acts as an internal ground. The second wire is a hot lead that provides a current signal when the ceramic core material changes its resistance with a change in temperature.

Another continuous-loop system, the Fenwal system(Figure 4-9-3), uses a single wire surrounded by a continuous string of ceramic beads in an inconel tube.

The beads in the Fenwal detector are wetted with a eutectic salt which possesses the characteristic of suddenly lowering its electrical resistance as the sensing element reaches its alarm temperature.

In both the Kidde and the Fenwal systems, the resistance of the ceramic or eutectic salt core meterial prevents electrical current from flowing at normal temperatures. In case of a fire or overheat condition, the core resistance drops and current flows between the signal wire and ground, energizing the alarm system.

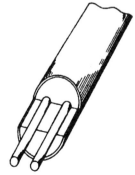

Figure 4-9-2 Kidde sensing

4-9. Warning and Fire Protection Systems

Overheat warning systems are used on some transport aircraft to indicate high temperature areas that could lead to a fire if a malfunction occurs. The number of overheat warning systems varies from aircraft to aircraft. On some aircraft, they are provided for each engine nacelle, wheel well area and for the area surrounding the pneumatic engine bleed air manifold. When an overheat condition occurs in the detector area, the systems causes a light on the fire control panels to warn the flight crew.

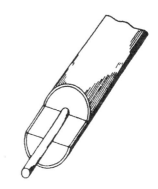

Figure 4-9-3 Fenwal sensing element

A smoke detection system is generally used to monitor the cargo and baggage compartments for the presence of smoke, which is indicative of a fire condition. Smoke detection instruments, which collect air for sampling, are mounted in the compartments in strategic location. A smoke detection systems is used where the type of fire anticipated is expected to generate a substantial amount of smoke before temperature changes are sufficient to actuate a heat detection system. Smoke detection instruments are classified by their method of detection.

One type of detector consists of a photoelectric cell, a beacon lamp, a test lamp, and a light trap, all mounted on a labyrinth. An accumulation of 10% smoke in the air causes the photoelectric cell to conduct electric current. Figure 4-9-4 shows the details of the smoke detector, and indicates how the smoke of the smoke detector, and indicates how the smoke particles refract the light to the photoelectric cell. When activated by smoke, the detector supplies a signal to the smoke detector amplifier. The amplifier signal activates a warning light and bell.

On some aircraft visual smoke detectors provide the only means of smoke detection. Indication is provided by drawing smoke through a line into the indicator, using either a

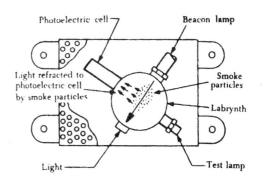

Figure 4-9-4 Photoelectric smoke detector

suitable suction device or cabin pressurization. When smoke is present, a lamp within the indicator is illuminated automatically by the smoke detector. The light is scattered so that the smoke is rendered visible in the appropriate window of the indicator. If no smoke is present, the lamp will not be illuminated. A switch is provided to illuminate the lamp for test purposes. A device is also provided in the indicator to show that the necessary airflow is passing through the indicator.

A. Types of Fires

The National Fire Protection Association has classified fires in three basic types. One type is the class A fire, defined as fire in ordinary combustible materials such as wood, cloth, paper, upholstery materials, etc. Class B fires are defined as fires in flammable petroleum products or other flammable or combustible liquids, greases, solvents, paints, etc. The third type of fire is the Class C fire which is defined as fire involving energized electrical equipment where the electrical non-conductivity of the extinguishing media is of great importance.

B. Aircraft Compartment Fire Classifications

The location of a fire is also identified by a letter designator. The cockpit and passenger cabin are designated Class A compartments, meaning that a fire may be visually detected, reached, and combatted by a crewmember. The engines are generally Class C compartments, and fire warning is provided by fire detectors. There are two basic types of cargo compartments, Class B, in which a crewmember may reach and combat a source of fire. The other type of compartment is, Class E or D, in which a crewmember can not reach the source of fire.

Compartments are classified Class A when they provide for visual detection of smoke and have an accessible inflight fire extinguisher available.

The cargo and baggage compartments are classified "B" if they have sufficient access to enable a member of the crew to move by hand all contents, and to reach effectively all parts of the compartment with a hand fire extinguisher, while in flight.

When the access provisions are being used, no hazardous quantity of smoke, flames, or extinguishing agent can enter any compartment occupied by the crew or passengers. Each compartment should be equipped with an approved type smoke detector or fire detector to give a warning to the flight deck. Hand fire extinguishers must be readily available for use in all compartments of this category.

4-9. Warning and Fire Protection Systems

Compartments are classified Class C when they have smoke or fire detectors installed and a built-in fire extinguisher system controlled from the cockpit.

Cargo and baggage compartments are classified "D" if they are so designed and constructed so that a fire occurring therein will be completely confined without endangering the safety of the airplane or the accupants. Ventilation and drafts controlled within each compartment must be configured so that any fire likely to occur in the compartment will not progress beyond safe limits. The compartment must be completely lined with fire resistant material.

On airplanes used to carry cargo only, the cabin area can be classified as a Class E compartment, if the window shades are closed and they are lined with fire-resistant material. A class E compartment must also be equipped with a separate system of an approved type smoke or fire detector with a means provided to shut off the ventilating air flow to, or within, the compartment. Controls for such means shall be accessible to the flight crew in the cockpit. A means of excluding hazardous quantities of smoke, flames or noxious gases from the cockpit must also be provided. The required crew emergency exits must be accessibel under all cargo loading conditions.

C. Extinguishing Agents

Aircraft fire extinguishing agents have some common characteristics which make them compatible to aircraft fire extinguishing systems. All agents must be able to be stored for long periods of time without adversely affectiong the system components or agent quality. The extinguishing agents must not freeze at normally expected atmospheric temperatures. The nature of the devices inside a powerplant compartment reuqire agents that are not only useful against flammable fluid fires, but also effective on electrically caused fires. Agents are classified into two general categories based on the mechanics of extinguishing action ; the halogenated hydrocarbon agents and the inert cold gas agents.

The probable extinguishing mechanism of halogenated agents is a chemical interference in the combustion process between fuel and oxidizer. Experimental evidence indicates that the most likely method of transferring energy in the combustion process is by molecule fragments resulting from the chemical reaction of the constituents. If these fragments are blocked from transferring their energy to the unburned fuel molecules, the combustion process may be slowed, or stopped completely(extinguished), It is believed that the halogenated agents react with the molecular fragments, thus preventing the energy transfer. This may be termed chemical cooling or energy transfer blocking. This extinguishing mechanism is much more effective than oxygen dilution and cooling.

Both carbon dioxide(CO_2) and Nitrogen(N_2) are effective cold gas extinguishing agents. Carbon dioxide, CO_2, has been used for many years to extinguish flammable fluid fires and fires involving electrical equipment. It is noncombustible and does not react with most substances It provides its own pressure for discharge from the storage vessel.

Normally, CO_2 is a gas, but it is easily liquefied by compression and cooling. After liquification, CO_2 will remain in a closed container as both liquid and gas. When CO_2 is then discharged to the atmosphere, most of the liquid expands to gas. Heat absorbed by the gas during vaporization cools and becomes a finely divided white solid, dry ice snow. CO_2 is about 1 1/2 times as heavy as air, which gives it the ability to replace air above buring surfaces and maintain a smothering atmosphere. CO_2 is effective as an extinguishant primarily because it dilutes the air and reduces the oxygen content so that the air will no longer support combustion. Nitrogen, N_2, is an even more effective extinguishing agent. Like CO_2, N_2 is an inert gas of low toxicity. N_2 extinguishes by oxygen dilution and smothering. Freon® gas is also used as a cold gas extinguishing agent.

D. Fire Extinguishing Systems

A high-rate-of-discharge system(HRD) provides high discharge rates through high pressurization, short feed lines, large discharge valves and outlets. Because the agent and pressurizing gas of a HRD system are released into the zone in one second or less, the zone is temporarily pressurized, and interrupts the ventilating air flow. The few, large sized outlets are carefully located to produce high velocity swirl effects for best distribution.

The fire protection system of most large transport aircraft consists of two subsystems ; a fire detection system and a fire extinguishing system. These two subsystems provide fire protection not only to the engine and nacelle areas but alwo to such areas as the baggage compartments and wheel wells as mentioned earlier.

The typical fire extinguishing portion of a complete fire protection system includes a cylinder, or cylinders, of extinguishing agent for each engine and nacelle area. This type of system uses an extinguishing agent container similar to the type shown in Figure 4-9-5.

The container is equipped with two discharge valves which are operated by electrically discharged cartridges. These two valves are the main and the reserve controls, which release and route the extinguishing agent to the engine. A pressure gauge, a discharge manifold, and a safety discharge connection are provided for each container as shown in Figure 4-9-5.

4-9. Warning and Fire Protection Systems

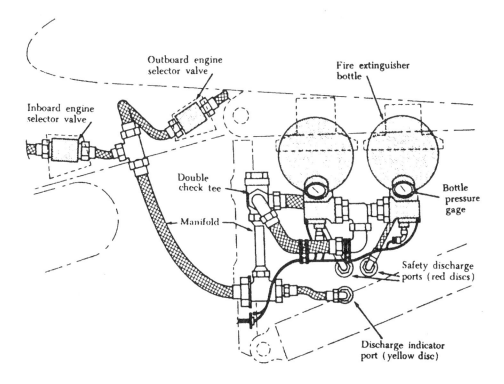

Figure 4-9-5 Dual container installation and fittings

The safety discharge connection is capped with a red indication disk. If the temperature rises beyond a predetermined safe value, the disk will rupture, dumping the agent overboard. A missing red disk in the safety discharge ports indicates a thermal(overpressure) discharge. A yellow disk discharge indicator port is used to indicate that the system was fired from the cockpit. If the system was fired(activated) by the flight crew, this yellow disk will be missing.

2) Lockheed L-1011 Fire Protection Systems

The L-1011 has both fire detection and fire extinguishing capabilities for the three engines and the APU, as can be seen in Figure 4-9-6. It has a fire(overheat) detection system for the main landing gear wheel wells, but no fire extinguishing capability in these

276 제4장 Aircraft System Description

locations. The galley is equipped with a smoke detection system, and the galley oven ventilation system is equipped with a duct overheat detector. Portable fire extinguishers are located at strategic points throughout the flight station, passenger compartments, and the galley.

Fire detection is accomplished by dual loop sensors located at strategic points on the engines, in the APU compartment, and in the main landing gear wheel wells. Each loop is a continuous system with the sections of the sensors connected to each other by aircraft wiring. The sensors are identified as loop A and loop B. A sensor section consists of both loops parallel to each other and attached to a support tube by quick-release clamps. A Teflon®-asbestos grommet around the sensor is located at each clamp. The sensor support tube is attached to the structure by a quick-release fastener as shown in Figure 4-9-7. The location of the engine fire detection sensors and interconnect wiring is illustrated in Figure 4-9-8.

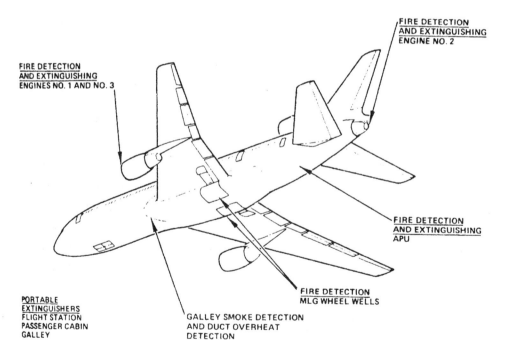

Figure 4-9-6 L-1011 Fire protection provistions(Lockheed)

4-9. Warning and Fire Protection Systems

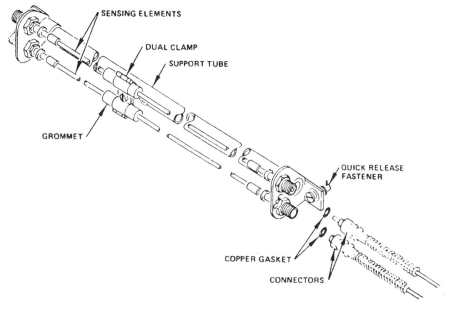

Figure 4-9-7 Dual fire detection sensors(Lockheed)

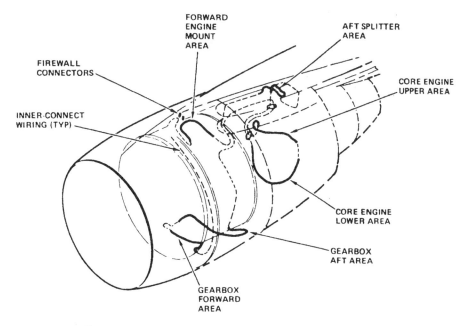

Figure 4-9-8 Engine fire detector sensors(Lockheed)

278 제4장 Aircraft System Description

A. Galley/Lounge Smoke Detector

The detector unit, illustrated in Figure 4-9-9, consists of a test lamp, a beacon lamp, a labyrinth, and a light sensitive resistor. The labyrinth is located between the beacon lamp and the light sensitive resistor so that very little light from the beacon lamp can reach the resistor during normal conditions. However, when smoke is drawn through the labyrinth, the smoke becomes a medium by reflecting or scattering the light so that more light can reach the resistor, reducing the resistance value of the light sensitive resistor. The light sensitive resistor is connected to the smoke warning circuits. When the value of the resistance is reduced sufficiently due to the increased light intensity, the smoke warning circuits will be activated.

During a smoke detection system test, the test lamp is connected in series with the beacon lamp, and light from the test lamp shines directly on the light sensitive resistor. Therefore, perfoming a system test checks the integrity of the smoke detector system and the continuity of the beacon lamp.

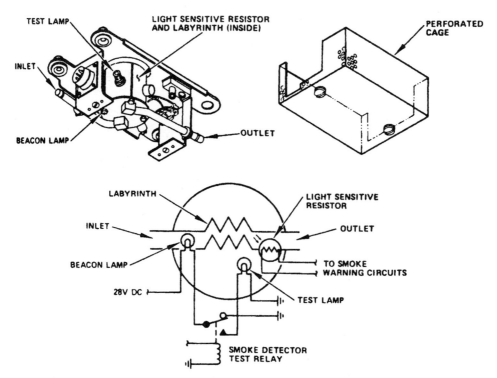

Figure 4-9-9 Galley/lounge smoke detector(Lockheed)

4-9. Warning and Fire Protection Systems 279

B. Fire Extinguisher System

The engines have two bottles located in the left hand side of the nose cowl of each pylon mounted engine, as illustrated in Figure 4-9-10. Bottle pressure gauges are mounted on each bottle and can be viewed through a Plexiglass window mounted in the bottle access panels. A thermal discharge indicator is located adjacent to the viewing windows. A 3/4-inch tube carries the fire extinguishing agent to a discharge nozzle located in the nose cowl to fan case mating point. The nozzle discharges the agent into the fan case and accessory sections of the engine.

The engine No.2/APU fire bottles serve both engine No.2 and/or the APU. Pressure gauges and thermal discharge indicators can be viewed from ground level in much the same manner as pylon mounted engines.

A tube carries the fire extinguishing agent to a discharge nozzle located in the forward end of the APU compartment. Another tube carries the fire extinguishing agent to two discharge nozzles for engine No.2. The engine No.2 nozzle are located on the engine No.2 firewall and discharge the fire extinguishing agent into the fan case and accessory section of the engine. The engine/APU fire bottles use Freon 1301, which is nitrogen pressurized.

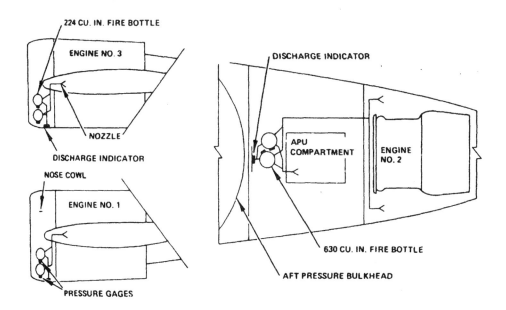

Figure 4-9-10 Fire extinguisher layout(Lockheed)

280 제4장 Aircraft System Description

C. Wing Engine Fire Extinguisher

The wing engine fire extinguisher bottles, shwon in Figure 4-9-11, are corrosion resistant steel spheres, with a volume of 224 cubic inches each. They contain 4.3 pounds of monobromotrifluoromethane and are charged with nitrogen to a pressure of 600±25PSI at 70 degrees F. Each bottle assembly includes a pressure gauge and pressure switch, and initiator cartridge, a discharge port, and a thermal discharge safety.

The pressure gauge indicates internal pressure of the bottle. A pressure switch closes when pressure decreases and illuminates the appropriate discharge light adjacent to the fire pull handle at the flight station. The pressure switch is an integral part of the pressure gauge and therefore cannot be replaced without replacing the complete bottle assembly. The pressure gauge reads from 0 to 1500PSI.

The fire extinguishing agent is released from the bottle by utilizing an initiator cartridge, or squib, to rupture a frangible disc. A strainer retains the pieces of the ruptured disc. A strainer retains the pieces of the ruptured disc and prevents them from entering the

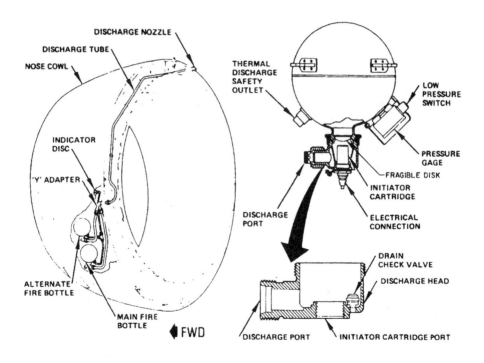

Figure 4-9-11 L-1011 Wing engine fire extinguisher installation(Lockheed)

4-9. Warning and Fire Protection Systems   281

discharge line. The initiator is an electrically fired cartridge.

Each cartridge assembly consists of two squibs, electrically connected in parallel. A DC voltage of 10-29 volts, at a minimum of 3 amps, is required to fire each squib. The cartridges are fired by actuation of a switch located adjacent to the applicable fire pull handle.

There is a drain check valve installed in each discharge head. The valve is spring loaded open which allows moisture to drain overboard and not accumulate around the initiator cartridge. This helps prevent misfire. When the bottle is fired, pressure buildup in the head will close the valve at 35PSI and prevent loss of agent.

The thermal discharge outlet is designed to open if the pressure in the bottle should exceed its maximum limit. Rupture of the thermal discharge will be indicated by a missing red thermal discharge disk indicator and by the appropriate DISCH(discharge) light in the flight station.

Engine No.2 and the APU share two fire bottles. Each bottle has two sets of initiator

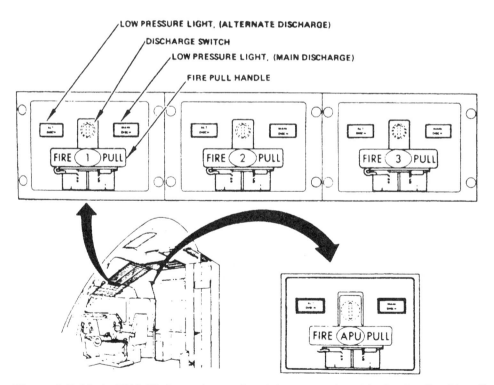

Figure 4-9-12 L-1011 Flight station extinguisher control and indication(Lockheed)

282 제4장 Aircraft System Description

cartridges. One cartridge, when fired, will direct the agent to engine No.2, and the other cartridge, when fired, directs it to the APU compartment. The bottles are larger than those mounted on the pylon engines.

A fire pull handle is provided for each of the three engines and the APU, as shown in Figure 4-9-12. Pulling the handle exposes a discharge switch. Moving the switch to the right discharges the main bottle and moving the switch to the left discharges the alternate bottle. When a bottle is discharged, the appropriate(MAIN or ALT) light adjacent to the fire pull handle is illuminated. Since the No.2 engine and the APU share the same bottles, both sets of discharge lights will be illuminated. If a fire bottle is discharged due to a bottle overpressure(thermal) condition, the DISCH light will also be illuminated.

The fire extinguisher test panel is located at the flight engineer station. Main and alternate test lights for each of the engines and the APU system determine the integrity of the electrical cartridge firing circuit during test. The test switch verifies continuity to the cartridges and the short switch checks the circuits for a short or grounded condition.

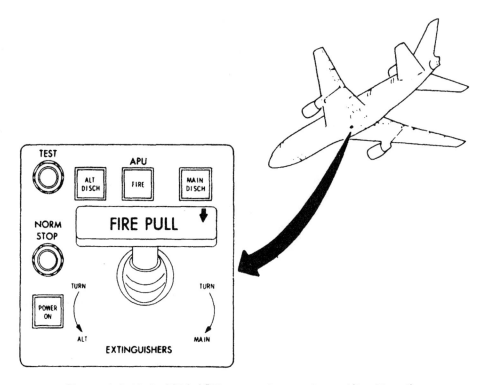

Figure 4-9-13 L-1011 APU external control panel(Lockheed)

4-9. Warning and Fire Protection Systems

The APU fire extinguishing system is similar to the engine extinguishing systems. The APU fire extinguishing system contains two additional discharge indicators and an additional fire handle on the APU external control panel, as illustrated in Figure 4-9-13. The APU extinguishing system also includes an auto APU extinguishing relay which, when energized, will automatically discharge the main fire bottle into the APU compartment.

3) Aircraft Warning Systems

Most aircraft warning systems are either visual and/or aural warnings. Steady or flashing lights and EICAS messages are used to alert the flight crew visually. Aural warnings can make several different sounds. For example, the aircraft overspeed system makes a clacking sound, while the unsafe takeoff or landing system uses a horn. Each different sound represents a particular warning. Waring systems, as they pertain to specific aircraft, will be discussed with each aircraft.

4) L-1011 Aural Warning Systems

The L-1011 uses several aural warning systems, as illustrated in Figure 4-9-14. The cabin pressure warning system is fairly typical of other aircraft in that at 10,000feet cabin altitude, an intermittent horn will sound to alert the flight crew of an unsafe cabin pressure. The aircraft's overspeed warning system is activated any time the aircraft's speed exceeds the maximum allowable airspeed. It will remain activated unitl the speed is reduced to safe limits. The aural warning system condition and sounds for other unsafe aircraft conditions are described in Figure 4-9-14. Other warning systems will be discussed in more detail in the section on Boeing 737-300.

5). Boeing 737-300 Fire and Warning Systems

A. Fire Protection
Fire protection consists of overheat and fire detection sensors and fire extinguishers. Detection provides visual and aural indications of overheat and fire conditions in the engines and main wheel well areas. The extinguishers provide a means of extinguishing engine and APU fires.

Four dual element overheat/fire detection loops are installed in each engine nacelle.

284 제4장 Aircraft System Description

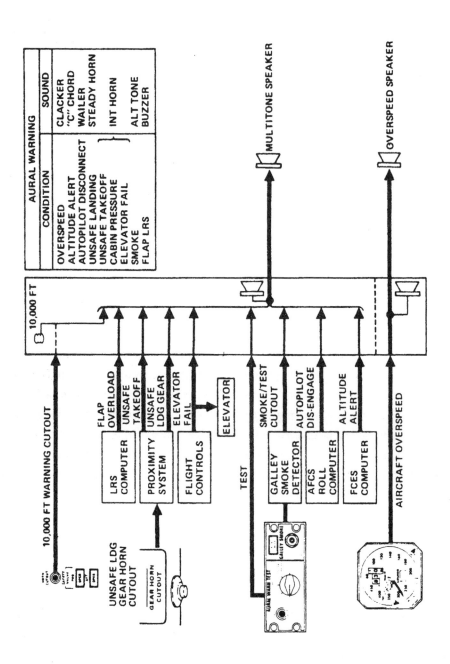

Figure 4-9-14 Aural warning system(Lockheed)

4-9. Warning and Fire Protection Systems

OVERHEAT DETECTOR SWITCH

NORMAL - Both the A loop and the B loop must sense an overheat or fire condition before a warning is activated.

A or B - Only the selected detector loop initiates an overheat or fire warning.

FIRE WARNING BELL CUTOUT SWITCH

PRESS - Silences the fire alarm bell and cancels the master FIRE WARNING lights.
- Silences the APU horn in the main wheel well.

EXTINGUISHER TEST SWITCH

1 and 2 - Tests the associated bottle discharge circuits for all three extinguisher bottles.
- All three extinguisher test lights illuminate in 1 and 2.

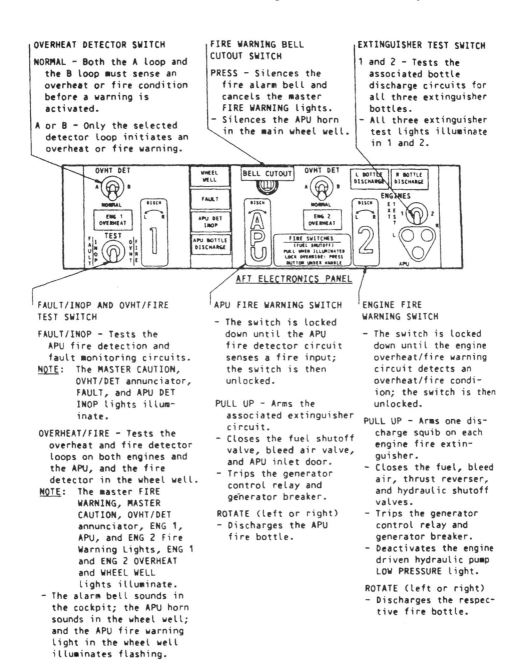

FAULT/INOP AND OVHT/FIRE TEST SWITCH

FAULT/INOP - Tests the APU fire detection and fault monitoring circuits.
NOTE: The MASTER CAUTION, OVHT/DET annunciator, FAULT, and APU DET INOP lights illuminate.

OVERHEAT/FIRE - Tests the overheat and fire detector loops on both engines and the APU, and the fire detector in the wheel well.
NOTE: The master FIRE WARNING, MASTER CAUTION, OVHT/DET annunciator, ENG 1, APU, and ENG 2 Fire Warning Lights, ENG 1 and ENG 2 OVERHEAT and WHEEL WELL lights illuminate.
- The alarm bell sounds in the cockpit; the APU horn sounds in the wheel well; and the APU fire warning light in the wheel well illuminates flashing.

APU FIRE WARNING SWITCH

- The switch is locked down until the APU fire detector circuit senses a fire input; the switch is then unlocked.

PULL UP - Arms the associated extinguisher circuit.
- Closes the fuel shutoff valve, bleed air valve, and APU inlet door.
- Trips the generator control relay and generator breaker.

ROTATE (left or right)
- Discharges the APU fire bottle.

ENGINE FIRE WARNING SWITCH

- The switch is locked down until the engine overheat/fire warning circuit detects an overheat/fire condition; the switch is then unlocked.

PULL UP - Arms one discharge squib on each engine fire extinguisher.
- Closes the fuel, bleed air, thrust reverser, and hydraulic shutoff valves.
- Trips the generator control relay and generator breaker.
- Deactivates the engine driven hydraulic pump LOW PRESSURE light.

ROTATE (left or right)
- Discharges the respective fire bottle.

Figure 4-9-15 Boeing 737 overheat/fire protection panel switches (Boeing)

The detectors are the Kidde sensor element type. At a predetermined temperature, the sensor activates the overheat warning system. At a higher temperature, the fire warning system is activated.

The dual element detectors are labeled A and B. An OVHT DET(overheat detection) switch for each engine, labeled A, B, and NORMAL, permits selection of loop A, B, or both A and B as the active detecting element(s). Normally operating as a dual-loop system, an alert is initiated only if both loops detect an overheat or fire condition.

An engine overheat condition is indicated by the illumination of the MASTER CAUTION light, OVHT/DET annunciator and the associated engine OVERHEAT light. The overheat light remains illuminated unitl the temperature drops below the onset temperature.

An engine fire condition is indicated by the illumination of the master FIRE WARNING and associated engine fire switch light and the sound of the alarm belll The bell may be silenced and the master FIRE WARNING lights extinguished by pressing either master FIRE WARNING light or the bell cutout switch on the fire panel, as shown in Figure 4-9-15.

B. Fire Detection

A Kidde sensor element fire detector loop is installed on the APU. At a predetermined alarm temperature, the sensor activates the warning signals.

If a fire is present, the FIRE WARNING lights illuminate, the bell sounds, the APU fire switch illuminates and the APU shuts down. The warning horn in the main wheel weell also sounds if the airplane is on the ground(see Figure 4-9-16). When the FIRE WARNING lights are illuminated, a fire is assumed and should be extinguished. The fire switch remains illuminated until the temperature surrounding the sensor/responder has decreased below the alarm temperature. Illumination of the amber APU DET INOP light, located on the fire panel, indicates a failure in the APU fire detector loop.

A fire detection loop is also installed in the main wheel well. The detector is a Fenwall metallic type. Testing the system checks the continuity of the loop by sending an artificial electronic signal to the fire warning system. The overheat and fire warning systems for the engines, APU, and wheel wells are shown in Figure 4-9-17.

The lavatory smoke detection system monitors air for the presence of smoke and provides an aural warning if smoke is detectred. Pressing the INTERRUPT switch silences the aural warning. If smoke is still present when the switch is released, the alarm will

4-9. Warning and Fire Protection Systems 287

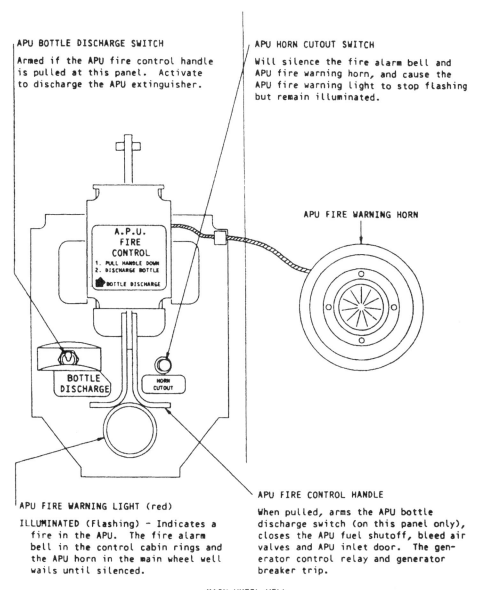

Figure 4-9-16 Boeing 737 APU ground control panel (Boeing)

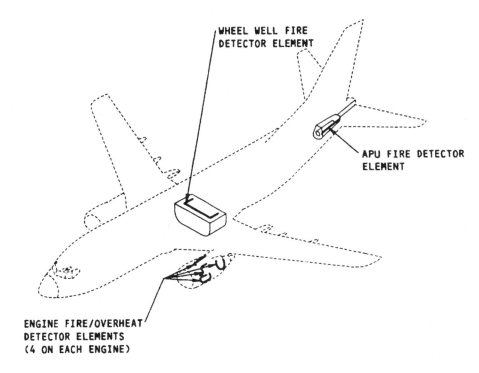

Figure 4-9-17 Fire and overheat detector element locations(Boeing)

sound again.

C. Engine Fire Extinguisher System

The engine fire extinguisher system(see Figure 4-9-18), consists of two Freon bottles with their associated plumbing to each engine, plus the fire warning switch, test, and bottle discharge lights.

The fire warning switch is normally locked down to prevent inadvertent shutdown of an engine. Illumination of an engine fire warning light, or engine overheat light, causes a solenoid to activate, which unlocks the fire warning switch. Pulling the engine fire warning switch up: arms one discharge squib on each engine fire extinguisher bottle; closes the fuel tank shutoff valve ; trips the generator control relay and breaker; and closes the hydraulic fluid shutoff valve. The engine driven hydraulic pump LOW PRESSURE light is deactivated. It also closes the engine bleed air valve resulting in loss of wing anti-ice to the affected wing and closure of the bleed air operated pack valve. Then the engine fire

4-9. Warning and Fire Protection Systems 289

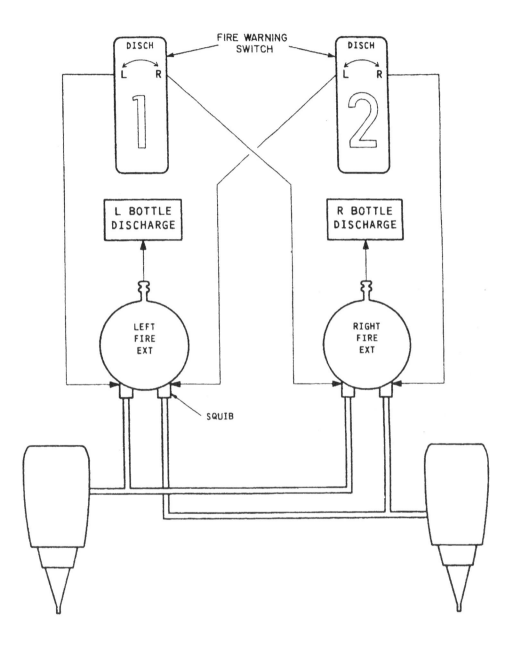

Figure 4-9-18 Engine fire extinguishing schematic(Boeing)

290 제4장 Aircraft System Description

warning switch can be rotated to discharge the fire extinguisher bottle.

Rotation of the engine fire warning switch electrically fires a squib, puncturing the seal of the extinguisher bottle, discharging the extinguishing agent into the associated engine. One or both bottles may be discharged into either engine. Rotating the switch the other way discharges the remaining bottle. A BOTTLE DISCHARGE light illuminates a few seconds after the fire warning switch is rotated, indicating the bottle has discharged.

D. APU Fire Extinguisher System

The APU fire extinguisher system consists of one Freon® bottle with its associated plumbing to the APU, plus a fire warning switch, test and bottle discharge light. Pulling the fire warning switch up, shown in Figure 4-9-15, provides backup for the automatic shutdown feature, arms the extinguisher circuit, shuts down the APU by deactivating the fuel solenoid, and closes the fuel tank shutoff valve. The APU bleed air valve and APU inlet door also close. The generator control relay and generator braker trip off line. The APU BOTTLE DISCHARGE light illuminates, indicating the bottle has discharged.

E. Lavatory Fire Extinguisher System

An automatic fire extinguisher system, shown in Figure 4-9-19, is located beneath the sink area in each lavatory. The extinguisher discharges non-toxic Freon® gas through one, or both, of two heat-activated nozzles. One nozzle discharges toward the towel disposal container; the other directly under the sink. The color of the nozzle tips will change to an aluminum color when the extinguisher has discharged.

A temperature-indicator placard is located on the inside of the access door below each sink. White dots on the placard will turn black when exposed to high temperatures. If an indicator has turned black, or a nozzle tip has changed color, it should be assumed that the extinguisher has discharged. An inspection for fire damage should be made, the extinguisher replaced, and the temperature-indicator placard replaced before the next flight.

F. Warning System

The aural and visual warnings alert the flight crew to conditions that require action or caution in the operation of the airplane. The character of the signal used varies, depending upon the degree of urgency or hazards involved. Aural, visual, and tactile signals are used singularly, or in combinations, to provide simultaneously, both warning and information regarding the nature of the condition.

4-9. Warning and Fire Protection Systems

Red warning lights located in the area of the pilots's primary forward field of vision are used to indicate engine, wheel well, or APU fires, autopilot disconnect(flashing), and landing gear unsafe conditions. Conditions which require timely corrective action by the flight crew are indicated by means of amber caution lights.

Various aural warnings call attention to warnings and cautions. An aural warning for airspeed limits is given by a clacker, the autopilot disconnect by a warning tone, cabin altitude by an intermittent horn, landing gear positions by a steady horn. The unsafe takeoff warning is given by an intermittent horn, and the fire warning by a fire warning bell. Ground proximity warnings and alerts are indicated by voice warnings. Generally, aurals automatically silence when the associated non-normal condition no longer exists.

G. Takeoff Configuration Warning

The takeoff configuration warning is armed when the airplane is on the ground and

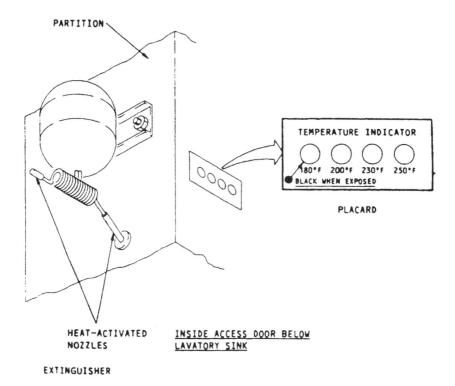

Figure 4-9-19 Lavatory fire extinguisher system (Boeing)

either, or both, forward thrust levers are advanced for takeoff. An intermittent takeoff warning horn sounds if: Stabilizer trim is not in the green band range, or; Trailing edge flaps are not in the takeoff range, or; Leading edge devices are not in the correct position for takeoff, or; Speed brake lever is not in the down(stowed) Position. The warning indication is cancelled when the configuration error is corrected.

H. Landing Gear Configuration Warnings

Visual indications and aural warnings of landing gear position are provided by the landing gear indicator lights and landing gear warning horn. The landing gear indication lights(visual indication) are activated by signals from each gear, the landing gear lever, and the forward thrust lever position switches. If the green light for each gear is illuminated, the landing gear is down and locked. When the red light is illuminated, the landing gear is in disagreement with the landing gear lever position(in transit or unsafe). When all of the lights are extinguished, the landing gear is in the up and locked position with the landing gear lever UP, or OFF.

A steady warning horn is provided to alert the pilots whenever the airplane is in a landing configuration and any gear is not down and locked. The landing gear warning horn is also activated by flap and thrust lever position.

Generally, when either thrust lever is retarded and the landing gear is in an unsafe condition, the landing gear warning horn will sound, but can be silenced using the warning horn cutout switch. Under certain conditions, the landing gear warning horn cannot be silenced. Although the actual flap settings and thrust lever positions will vary from one aircraft type to another, generally some provision is made to deactivate the horn cutout switch when the aircraft is in an actual landing configuration and the landing gear is not down and locked.

I. Mach/Airspeed Warning System

Two independent mach/airspeed warning systems, shown in Figure 4-9-20, provide a distinct aural warning any time the maximum operating airspeed is exceeded. The warning clackers can be silenced only by reducing airspeed below maximum operating speed.

·The systems operate from a mechanism internal to each pilot's mach/airspeed indicator. Test switches allow a system operation check at any time. Maximum operating airspeeds exist primarily due to airplane structural limitations at lower altitudes and airplane handling characteristics at higher altitudes.

4-9. Warning and Fire Protection Systems 293

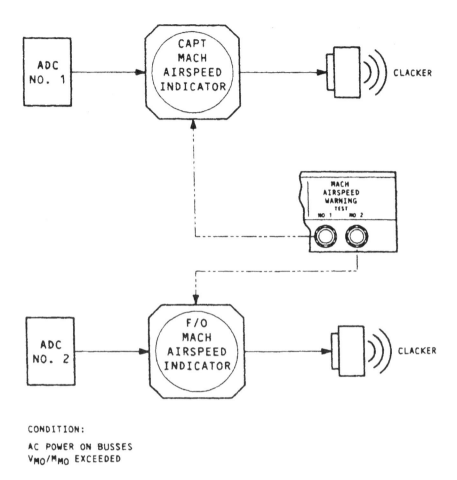

CONDITION:
AC POWER ON BUSSES
V_{MO}/M_{MO} EXCEEDED

figure 4-9-20 Mach/airspeed warning schematic(Boeing)

J. Stall Warning System

Warning of an impending stall is required to occur a minimum of seven percent above actual stall speed. Natural stall warning(buffet) usually occurs at a speed prior to stall. In some configurations, the margin between stall and stall warning(buffet) is less than the requried seven percent. Therefore, an artificial stall warning device, a stick shaker, is

294 제4장 Aircraft System Description

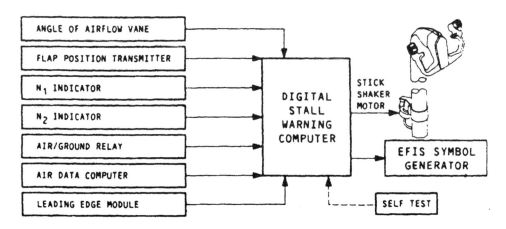

Figure 4-9-21 Boeing 737 stall warning system(Boeing)

utilized to provide the required warning.

The stall warning, or stick shaker, shown in Figure 4-9-21, consists of two eccentric weight motors, one on each control column. It is designed to alert the pilots before a stall develops. The warning is given by vibrating both control columns. The system is energized in flight at all times. The system is deactivated on the ground by the air/ground safety sensor.

Two independent digital computers are installed which compute the proper stall warning based on angle of attack, flap configuration and thrust. In addition, the stall warning occurs when the airspeed falls below a specified value for the selected flap setting, this is referred to as the speed floor. The computers receive inputs from the angle of airflow vanes, the flap position transmitter, the N1 and N2 indicators, the air/ground relay, the air data computers, and the leading edge module, as can be seen in Figure 4-9-21.

K. Ground Proximity Warning System(GPWS)

The ground proximity warning system(GPWS) provides warnings and/or alerts to the flight crew when one of the following conditions exist:
- Mode 1 — Excessive descent rate
- Mode 2 — Excessive terrain closure rate
- Mode 3 — Altitude loss after takeoff or go-around
- Mode 4 — Unsafe terrain clearance when not in the landing configuration

4-9. Warning and Fire Protection Systems

- Mode 5 — Excessive deviation below an ILS(Instrument Landing System) glide slope
- Mode 6 — Descent below the selected minimum radio altitude
- Mode 7 — Windshear condition encountered

Warnings for Modes 1 and 2 consist of red PULL UP lights and the aural "WHOOP WHOOP PULL UP".

Alerts for Modes 1, 2, 3, and 4 consist of the red PULL UP lights and one of the following aurals : "SINK RATE", "TERRAIN", "DONT SINK", "TOO LOW GEAR", "TOO LOW FLAPS", OR "TOO LOW TERRAIN".

Alerting for Mode 5 consists of amber BELOW G/S(Glide Slope) lights and the aural "GLIDE SLOPE".

Windshear warnings consist of red WINDSHEAR lights and a siren followed by the aural "WINDSHEAR".

Windshear warnings take priority over all other modes. A warnings and/or alert continues until the flight condition(s) is corrected. The ground proximity warning system adjusts the warning and alert envelopes to avoid nuisance warnings or alerts at airports with unique terrain conditions.

Mode 1 has two boundaries(see Figure 4-9-22) and is independent of airplane

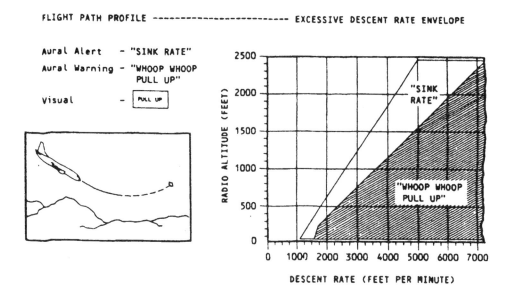

Figure 4-9-22 Mode 1 — Excessive descent rate(Boeing)

296 제4장 Aircraft System Description

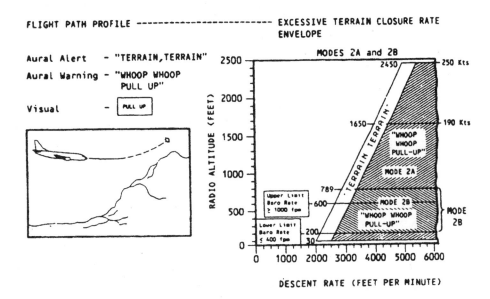

Figure 4-9-23 Mode 2 — Excessive terrain closure rate(Boeing)

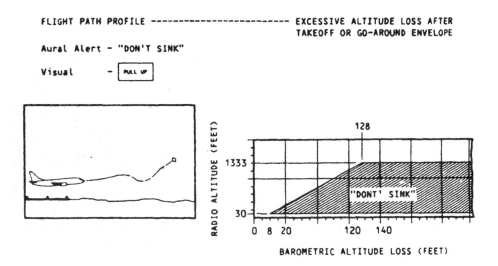

Figure 4-9-24 Mode 3 — Altitude loss after takeoff or go-around(Boeing)

4-9. Warning and Fire Protection Systems

configuration. Penetration of the first boundary activates the PULL UP lights and generates a repeated aural alert of "SINK RATE". Penetrating the second boundary causes the repeated aural warning of "WHOOP WHOOP PULL UP".

Mode 2 monitors airspeed, radio altitude and radio altitude rate of change, barometric altitude rate of change, and airplane configuration. Mode 2 has two boundaries. Penetrating the first boundary causes an aural alert of "TERRAIN" repeated twice, followed by the repeated aural warning "WHOOP WHOOP PULL UP". After leaving the PULL UP area, the repeating "TERRAIN" message will again be heard while in the terrain portion of the envelope(see Figure 4-9-23)

Mode 3(see Figure 4-9-24) provides an alert if a descent is made during the initial takeoff climb or during a go-around. Entering the envelope causes a repeated aural alert of "DON'T SINK". The alert continues unitl a positive rate of climb is established. If the airplane descends again before climbing to the original descent altitude, another alert is generated based on the original descent altitude.

The unsafe terrain clearnace mode(4A) with gear retracted is armed after takeoff upon climbing though 700 feet radio altitude. When this envelope is penetrated below 190 knots, the aural alert "TOO LOW GEAR" is sounded. Above 190 knots, the aural alert "TOO LOW TERRAIN" sounds.

Mode 4B provides an alert when the landing gear is down and flaps are not in the landing position.

When the envelope is penetrated below 159 knots, the aural alert "TOO LOW FLAPS" is repeated. Above 159 knots, the aural alert "TOO LOW TERRAIN" sounds.

Mode 5 alerts the flight crew of an excessive descent as shown in Figure 4-9-25. The envelope has two areas of alerting, soft and loud. In both areas, the alert is a repeated voice message of "GLIDE SLOPE" and illumination of both pilots' BELOW G/S lights. The voice message amplitude is increased when entering the loud alerting area. In both areas, the voice message repetition rate is increased as glide slope deviation increases and radio altitude decreases.

Mode 6 is operated between 50 and 1000 feet radio altitude. It provides an aural alert as the airplane descends through the minimum decision altitude(MDA) set on the capatin's radio altimeter. The alert is aural only and consists of "MINIMUMS, MINIMUMS" sounded one time.

The GPWS(ground proximity warning system) provides aural and visual warnings of windshear conditions(Mode 7). The aural warning consists of a two-tone siren followed by

298 제4장 Aircraft System Description

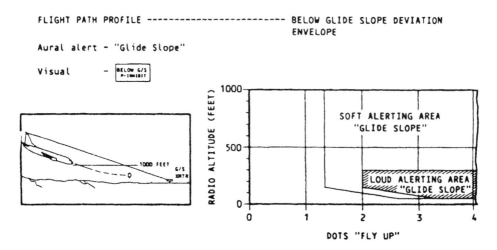

Figure 4-9-25 Mode 5 — Below glide slope deviation(Boeing)

the words "WINDSHEAR WINDSHEAR". The aural warning is activated only once during a windshear encounter. The visual warning is provided by illumination of the WINDSHEAR lights on the captain's and first officer's instrument panels. The light remains illuminated until a safe airspeed has been re-established after the windshear has dissipated.

4-10. Communications, Instruments, and Navigational Systems

1) Communications

External communication systems primarily involve voice transmission and reception between aircraft or aircraft and ground stations. some of the external communication systems that are used on large aircraft are high frequency system(HF), very high frequency systems(VHF), and selective calling(SELCAL) systems.

Internal communication systems, or systems used to communicate within or around the aircraft, can consist of a flight interphone,service interphone, cabin interphone, and passenger address system,. Each interphone system is similar in operation and will be described later in this chapter.

Another communication system, the aircraft communications addressing and reporting system(ACARS) is a data-link system which sends information between an aircraft and the airline ground base.

Some of the communication applications of ACARS that are common to many airlines are :
- Crew identification
- Out of gate time
- Off the ground time
- On the ground time
- In the gate time
- Dispatch and weather updates
- Engine performance
- Fuel status
- Passenger services
- Maintenance information
- ATIS(automatic terminal information service)

Radios are also used as navigational aids in a number of applications. They range from a simple radio direction finder, to navigational systems which use computers and other advanced electronic techniques to automatically solve the navigational problems for an entire flight.

Marker beacon receivers, instrument landing systems, distance measuring equipment, radar, area navigation systems, and omnidirectional radio receivers are but a few basic

300   제4장 Aircraft System Description

applications of airborne radio navigation systems available for use in aircraft. Other systems used to navigate and control the aircraft in flight can include autopilots(automatic flight), flight directors, flight management, and autothrottle systems.

The most common communication system in use is the VHF(Very high frequency) system. In addition to VHF equipment, large long range aircraft are usually equipped with HF(high frequency) communication systems. VHF airborne communication sets operate in the frequency range from 108.0 MHz to 135.95 MHz.

In general, the VHF radio waves follow approximately straight lines. Theoretically, the range of contact is the distance to the horizon and this distance is determined by the heights of the transmitting and receiving antennas. However, comunication is sometimes possible many hundreds of miles beyond the assumed horizon range.

A high frequency communication system(Figure 4-10-1) is used for long-range communication. HF systems operate essentially the same as a VHF system, but operate in the frequency range from 3 MHz to 30 MHz. Communications over long distances are possible with HF radio because of the longer transmission range. HF transmitters have higher power outputs than VHF transmitters.

A. Aircraft Communication Addressing and Reporting System(ACARS)

ACARS messages, mentioned earlier, are sent from the airplane via a VHF communication transceiver and a ground network to the airline ground operations base, and vice versa.

There is nothing new about sending messages between the airplane and the ground. What makes ACARS unique is that messages concerning everything from the contents of the fuel tanks and maintenance problems to food and liquor supplies can be sent by ACARS in a fraction of the time it takes using voice communications, in many cases without involving the flight crew.

Each ACARS message is compressed and takes about one second of air time to transmit. Sending and receiving data over the ACARS network reduces the number of voice contacts required on any one flight, thereby reducing communication workload. ACARS messages are limited to a length of 220 characters which is adequate for routine messages. Longer messages, known as multi-block messages, can be sent as a series of separate ACARS messages.

The ACARS network is made up of three sections: the airborne system; the service provider ground network; and the airline operations center(see Figure 4-10-2). The

4-10. Comm. Inst. and Nav. Systems 301

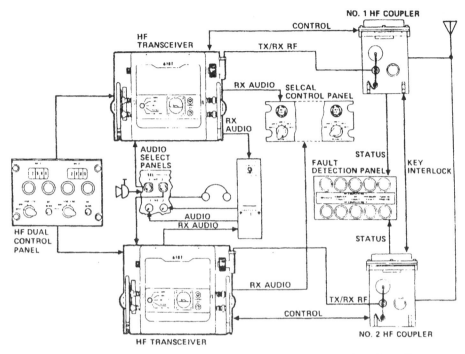

Figure 4-10-1 L-1011 HF communication system (Lockheed)

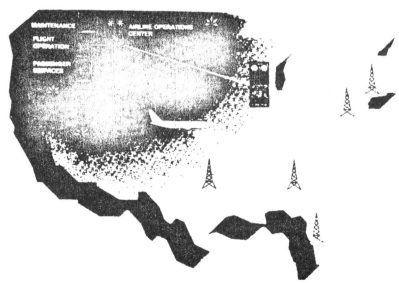

Figure 4-10-2 Components of the ACARS Network (Boeing)

airborne system(see Figure 4-10-3) has an ACARS management unit(MU) which manages the incoming and outgoing messages, and a control unit(CU) which is used by the crew to interface with the system. A printer may be installed to print incoming messages.

ACARS may be connected to other airplane systems such as the digital flight data acquisition unit(DFDAU). The DFDAU collects data from many of the airplane systems such as air data, navigation, and engine instruments, and in turn makes the data available to the ACARS. More recent ACARS installations have been connected to the flight management computer, permitting flight plan updates, predicted wind data, takeoff data and position reports to be sent over the ACARS network.

The ACARS ground system is made up of two parts(see Figure 4-10-2). The first part is the radio and message handling network, which is controlled in the United States by Aeronautical Radio Incorporated(ARINC). The service provider operates the remote radio sites, the ACARS front end processing system and the electronic switching system act like a post office and ensure that messages are routed to the correct addressee.

All ACARS messages originated in the United States are relayed from the remote sites at which they are received to the message handling center in Chicago. The ground systems in other parts of the world work in a similar way to the US system.

The third part of the network is an airline's operations or message center. The ARINC network is connected to the airline operations center by a land-line. At the airline, the message handling is performed by a computer system which sends received messages to the appropriate department(operations, engineering, maintenance, customer services) for the appropriate action. Messages from an airline department, such as a request from engineering for engine data, follow the same route in reverse.

The ACARS in use vary greatly from one airline to another and are tailored to meet each airline's operational needs. When satellite communication system are adopted in the near future. ACARS will take on a truly global aspect.

B. SELCAL

The SELCAL decoder is designed to relieve the flight crew from continuously monitoring the aircraft radio receivers. The SELCAL decoder is in effect an automatic monitor which listens for a particular combination of tones which are assigned to the individual aircraft by ARINC. Whenever a properly coded transmission is received from a ground ARINC station, the signal is decoded by the SELCAL unit, which then gives a signal to the flight crew indicating that a radio transmission is being directed to the aircraft.

4-10. Comm, Inst, and Nav. Systems   303

The flight crew can then listen to the appropriate receiver and hear the message.

ARINC ground stations equipped with tone-transmitting equipment call individual aircraft by transmitting two pairs of tones which will key only an airborne decoder set to respond to the particular combination of tones. When the proper tones are received, the decoder operates an external alarm circuit to produce a chime, light, buzz, or a combination of such signals.

A ground operator who wishes to contact a particular aircraft by means of the SELCAL unit selects the four-tone code which has been assigned to the aircraft. The tone code is transmitted by a radio-frequency wave, and the signal can be picked up by all receivers tuned to the frequency used by the transmitter. The only receiver which will respond to the signal and produce the alert signal for the flight crew is the SELCAL decoder system which has been set for the particular combination of tone frequencies generated by the ground operator.

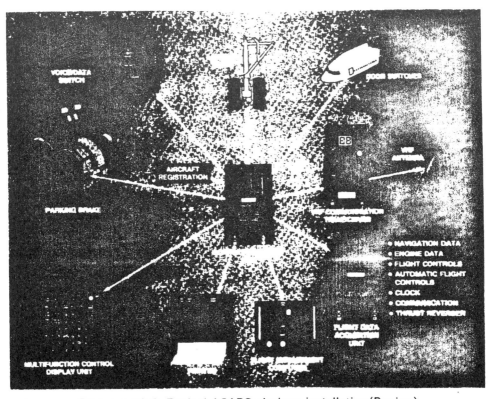

Figure 4-10-3 Typical ACARS airplane installation (Boeing)

2) Navigation Equipment

"Airborne navigation equipment" is a phrase which describes many different systems. The VHF VOR(omnidirectional range) is an electronic navigation system. As the name implies, the omnidirectional, or all-directional range station, provides the pilot with courses from any point within its service range. It produces 360 usable radials or courses, any one of which is a radio path connected to the station. The radials can be considered as lines that extend from the transmitter antenna like spokes of a wheel.

Operation is in the VHF portion of the radio spectrum(frequency range of 108.0MHz - 117.95MHz) with the result that interference from atmospheric and precipitation static is negligible. The navigational information is visually displayed on an instrument in the cockpit.

A. Instrument Landing System

The ILS(instrument landing system), can be visualized as a slide made of radio signals on which the aircraft can be brought safely to the runway. The entire system consists of a runway localizer, a glide slope signal, and marker beacons for position location. The localizer equipment produces a radio course aligned with the center of an airport runway.

The glide slope is a radio beam which provides vertical guidance to the pilot, assisting him in making the correct angle of descent to the runway. Glide slope signals are radiated from two antennas located adjacent to the touchdown point of the runway, as shown in Figure 4-10-4.

Two antennas are usually required for ILS operation. One for the localizer receiver, also used for VOR navigation, and one for the glide slope. The glide slope antenna is generally located on the nose area of the aircraft.

B. Marker Beacons and Distance-Measuring Equipment

Marker beacons are used in connection with the instrument landing system. The markers are signals which indicate the position of the aircraft along the approach to the runway. Two markers are used in each installation. The location of each marker is identified by both an aural tone and a signal light.

The purpose of DME(distance-measuring equipment) is to provide a constant visual indication of the distance the aircraft is from a ground station. The aircraft is equipped with a DME transceiver which is tuned to a selected DME ground station. Usually DME ground

4-10. Comm. Inst. and Nav. Systems   305

stations are located in conjunction with a VOR facility(called VORTAC).

The airborne transceiver transmits a pair of spaced pulses to the ground station. The pulse spacing serves to identify the signal as a valid DME interrogation. After reception of the challenging pulses, the ground station responds with a pulse transmission on a separate frequency to send a reply to the aircraft. Upon reception of the signal by the airborne transceiver, the elapsed time between the challenges and the reply is measured. This time interval is a measure of the distance separating the aircraft and the ground station. This distance is indicated in nautical miles.

C. Automatic Direction Finders

ADF(automatic direction finders) are radio receivers equipped with directional antennas which are used to determine the direction from which signals are received. Most ADF receivers provide controls for manual operation in addition to automatic direction finding.

When an aircraft is within reception range of a radio station, the ADF equipemnt provides a means of fixing the position with reasonable accuracy. The ADF operates in the low and medium frequency spectrum from 190 kHz through 1,750 kHz. The direction to the station is displayed on an indicator as a relative bearing to the station.

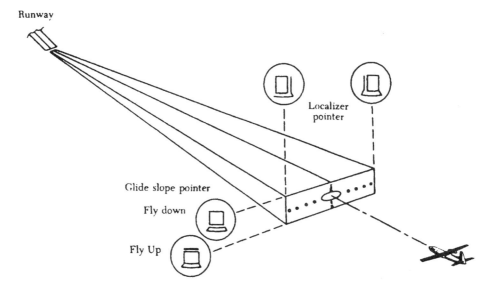

Figure 4-10-4 Localizer and glide slope information

D. Radar Beacon Transponder(ATC Transponder)

The radar beacon transponder system is used in conjunction with a ground base surveillance radar to provide positive aircraft identification and location directly on the controller's radar scope.

The airborne equipment, or transponder, receives a ground radar interrogation for each sweep of the surveillance radar antenna and automatically dispatches a coded response. The flight identification code, a four-digit number, is assigned during the flight planning procedure, or by air traffic control.

Most large airecraft transponders are equipped with an altitude encoding feature. The aircraft's altitude is transmitted to the ground station through the transponder.

E. Inertial Navigation System

The inertial navigation system is presently being used on large aircraft as a long-range navigation aid. It is a self-contained system and does not require signal inputs from ground navigational facilities. The system derives attitude, velocity, and heading information from measurement of the aircraft's accelerations.

Two accelerometers are required, one referenced to north and the other to east. The accelerometers are mounted on a gyrostabilized unit, called the stable platform, to avert the introduction of errors resulting from the acceleration due to gravity.

F. Airborne Weather Radar System

Rader(radio detection and ranging) is a device used to see certain objects in darkness, fog, or storms, as well as in clear weather. In addition to the appearance of these objects on the radar scope, their range and relative position are also indicated.

Radar is an electronic system using a pulse transmission of radio energy to receive a reflected signal from a target. The received signal is known as an echo, the time between the transmitted pulse and received echo is computed electronically and is displayed on the radar scope in terms of nautical miles. An L-1011 radar system is shown in Figure 4-10-5.

The weather radar increases safety in flight by enabling the operator to detect storms in the flight path and chart a course around them.

The terrain-mapping facilities of the radar show shorelines, islands, and other topographical features along the flight path. These indications are presented on the visual indicator in range and azimuth relative to the heading of the aircraft.

4-10. Comm. Inst. and Nav. Systems 307

G. Radio Altimeter

Radio altimeters are used to measure the distance from the aircraft to the ground. The indicating instrument will indicate the true altitude of the aircraft, which is its height above water, mountains, buildings, or other objects on the surface of the earth.

Radio altimeters are primarily used during landing. The altimeter provides the pilot with the altitude determine the decision flying the aircraft makes a decision whether to continue to land, execute a climb-out, or go-around.

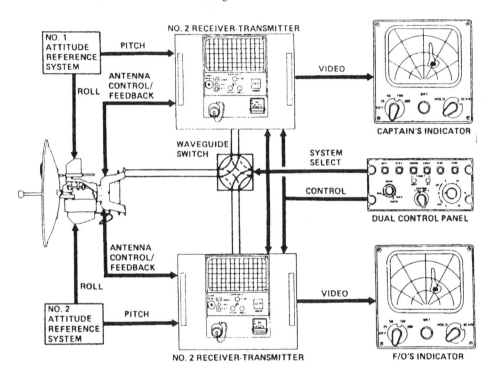

Figure 4-10-5 L-1011 Weather radar system(Lockheed)

3) Boeing 757 Avionics Systems

Avionics on the 757 airplane consist of communication, navigation, instrument, and autoflight systems. These systems allow the flight crew to communicate, navigate, control the airplane and manage the flight profile in the most efficient and effective manner possible.

308   제4장 Aircraft System Description

A. Communication

Communication equipment aboard the airplane allows the crew members to communicate with other aircraft, ground stations, other crew members, maintenance crews, and the passengers. The selective call system advises pilots of radio communication from another aircraft or a ground station and a voice recorder maintains a record of flight communication as mentioned earlier.

Three VHF communications radios are installed; designated L(left), C(center) and R(right). Control panels for the left and right are provided in the cockpit. Each panel allows the selection of two independent frequencies, the desired frequency being selected with the frequency transfer switch. Individual transmit and audio control is provided through the respective ASP(audio select panel).

An audio selector panel(ASP) is installed at the captain, first officer, and first observer stations. The panel allows each crew member to manage his own communications requirements.

The ASP permits two way(transmit and receive) communication capability from the flight interphone, cabin/service interphone, passenger address, and VHF radio systems. It provides voice and identification monitoring of selected navigation aids. Receiver volumes and microphone selections are also controlled from this panel. A cockpit loudspeaker, and volume control, is provided adjacent to each pilot seat.

B. Interphone Systems

There are two independent interphone systems in the aircraft, the flight interphone system and the cabin/service interphone system.

The flight interphone system(FIS) permits intercommunication between cockpit crew members without intrusion from the other phone system. Ground personnel are able to communicate on the FIS through a jack located on the APU ground control panel, shown in Figure 4-10-6.

The cabin interphone system(CIS) permits intercommunication between the cockpit and flight attendant stations. cockpit crewmembers communicate on the CIS through their audio select panel. The flight attendants communicate between flight attendant stations or with the cockpit using any of the handests in the cabin. The system is a party line similar to the 737 system. The service interphone system consists of additional internal and external jacks connected to the cabin interphone system for use by maintenance personnel.

The call system, illustrated in Figure 4-10-7, allows the cockpit crew, flight attendants

4-10. Comm, Inst, and Nav. Systems 309

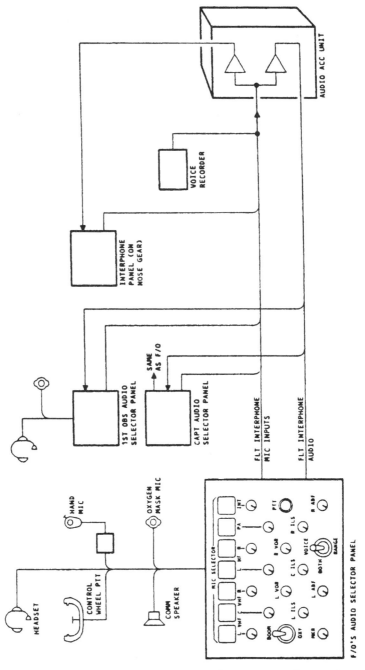

Figure 4-10-6 Boeing 757 flight interphone system (Boeing)

310 제4장 Aircraft System Description

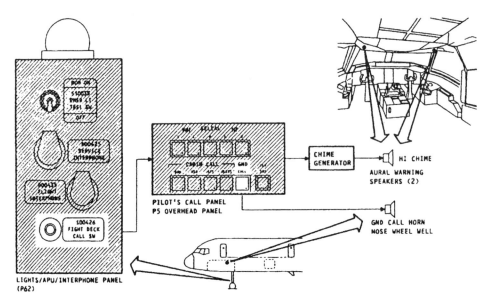

Figure 4-10-7 Boeing 757 crew call system(Boeing)

and ground personnel to indicate that interphone communication is desired. The cockpit crew can initiate calls through the pilot's call panel and are alerted through call lights and chimes.

The passenger address(PA) system, shown in Figure 4-10-8, allow cockpit crew and cabin attendants to make announcements throughout the cabin. Cockpit crewmembers can make announcements through any microphone and respective ASP. The passenger entertainment system(PES) consists of a multi-track tape player providing boarding music. PA announcements from any station override all tape player outputs.

C. Voice Recorder

The cockpit voice recorder records any transmissions from the cockpit made through the audio selector panels. it also records cockpit area conversations using an area microphone.

D. Navigational and Electronic Instrument Systems

Navigation equipment aboard the airplane provides the pilot information on flight conditions, aircraft position, performance, guidance, and flight profile. This information

4-10. Comm, Inst, and Nav. Systems 311

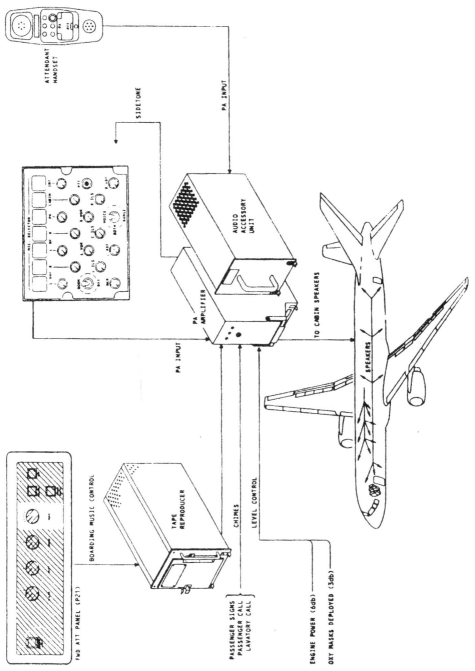

Figure 4-10-8 Boeing 757 passenger address system (Boeing)

312 제4장 Aircraft System Description

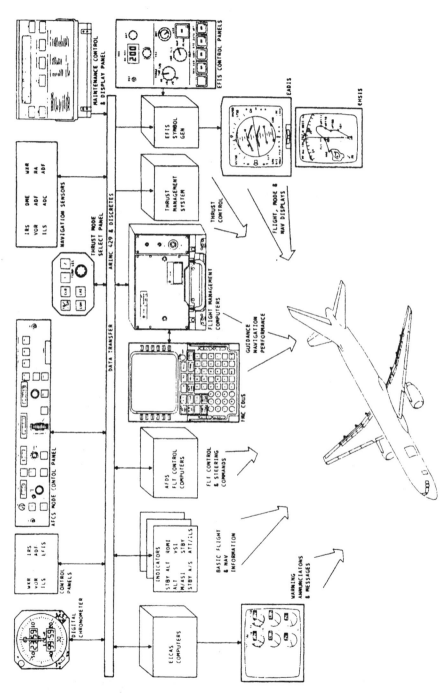

Figure 4-10-9 Boeing 757 navigation system(Boeing)

4-10. Comm. Inst. and Nav. Systems   313

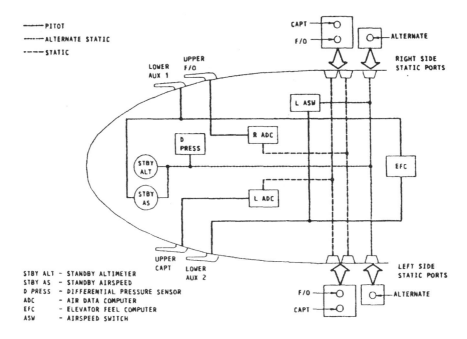

Figure 4-10-10 Boeing 757 pilot-static system (Boeing)

may be displayed directly or may be used as inputs to the flight management system, which then calculates a 3 dimensional flight path and steers the airplane along a selected route.

The navigation equipment is integrated with computer, display and cautionary systems into a flight management systems, illustrated in Figure 4-10-9. The majority of the units are digital, with microprocessors and software programs replacing many of the discrete components of previous units.

The instrument systems provide the pilot with primary flight information, selectable navigation displays, system mode annunciations, advisory, caution, and warning messages.

The pitot-static system, shown in Figure 4-10-10, senses both dynamic and static air pressures and supplies these pressures to the left and right air data computers, to the standby instruments, and to other airplane systems as required.

The pitot probes are located on the right and left sides of the forward fuselage body. The static ports are located on each side of the forward body. The pitot probes sense total air pressure while the static ports sense ambient static air pressure. These pressure are used

314 제4장 Aircraft System Description

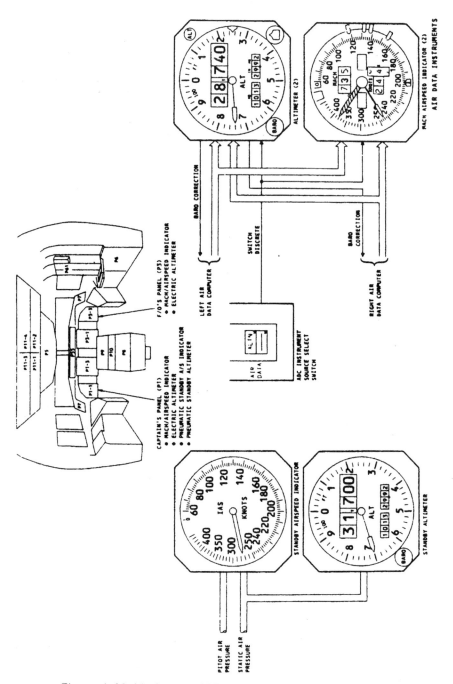

Figure 4-10-11 Boeing 757 air data instruments(Boeing)

4-10. Comm, Inst, and Nav. Systems   315

by the air data computer to determine the airplane's altitude, airspeed, and other related parameters.

The air data computer system monitors the airspace adjacent to the airplane and provides a single point source for altitude, airspeed, pressure, angle of attack and temperature data.

Inputs to the air data computers are total and ambient air pressure from the pitot static system, and barometric correction from the altimeters. Other inputs include total air temperature(TAT) from the TAT probe and angle of attack(AOA) from the AOA sensors.

The air data computers convert these analog inputs to digital signals and uses them to determine output values. The computer output signals are transmitted on four digital data buses which supply data to the air data instruments, engine and flight controls, navigation, warning, flight managment, and autoflight systems.

The air data instruments, shown if Figure 4-10-11, provide flight deck displays of altitude and airspeed. The standby airspeed indicator and altimeter are pneumatic instruments supplied with static and pitot air pressure from the pitot-static system. The

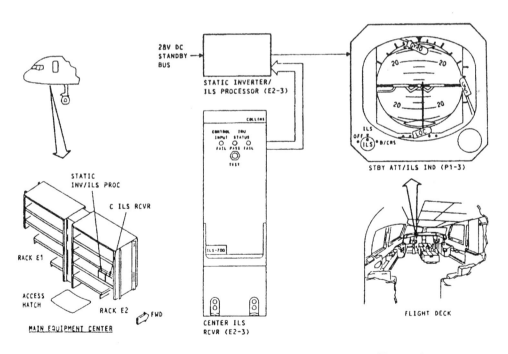

Figure 4-10-12 Standby attitude reference system(Boeing)

electric altimeters and mach airspeed indicators are supplied data from both the left and right air data computers.

The standby attitude reference system(see Figure 4-10-12) provides pitch and roll attitude information to the pilot in the event of failure of the two primary attitude displays. Localizer and glideslope information from the center ILS receiver is also displayed on the standby attitude indicator for use in the event of failure of the primary indicating systems. The attitude information is independent of other aircraft systems and requires only standby power.

The inertial reference system(IRS) senses airplane displacement about 3 axes to provide primary attitude, true and magnetic heading, vertical speed, navigation position, accelerations and angular rates, wind velocity and direction, and ground track. Each IRS contains 3 laser gyros, 3 accelerometers, power supplies, a microprocessor, BITE, and output circuitry. Each of the three inertial reference systems operates independently of the other two systems.

The IRSs receive barometric altitude, altitude rate, and true airspeed data from the air data computers. This information is used along with gyro and accelerometer data to determine the airplane's vertical speed and calculate wind parameters.

The air traffic control system(transponder) allows ground stations to monitor the aircraft's position continuously, by providing altitude and identity information.

Two separate systems(left and right) are installed. Each system consists of an antenna and transponder(transmitter/receiver) and are controlled from a single control panel. The control panel is used to set the assigned code, select the left or right system select altitude reporting, and initiate an identity pulse

The distance measuring equipment(DME), see Figure 4-10-13, measures slant range (DME distance) from the aircraft to selected DME ground stations and provides data for DME distance displays and continuous distance information to the flight management computers(FMCs). Two separate systems are installed, with each system consisting of one antenna and one interrogator. Three control panels are shared with the VOR and ILS receivers. The interrogator sends out interrogation pulse pairs to the selected ground station, which are then returned to the aircraft. The slant range is computed and displayed on the RDMIs and EHSIs, and is also routed to other navigation data users.

The automatic direction finder(ADF) provides the relative bearing of ground stations with repect to the airplane's heading. Standard broadcast stations and non-directional homing beacons are normally used with the ADF system.

4-10. Comm. Inst. and Nav. Systems 317

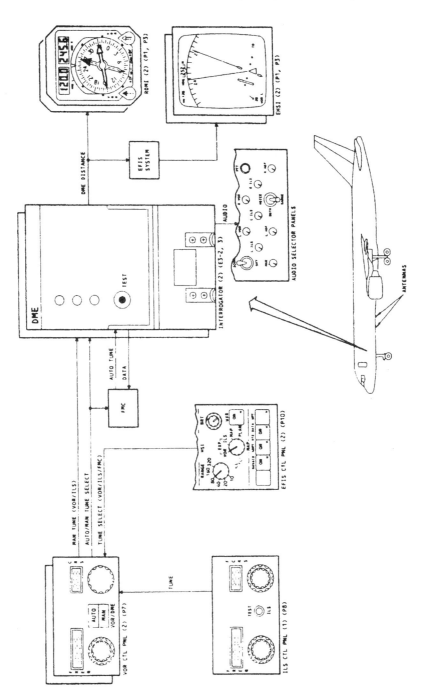

Figure 4-10-13 Boeing 757 distance measuring equipment(Boeing)

The VOR navigation system, illustrated in Figure 4-10-14, provides magnetic bearing to or from a ground VOR transmitter. Two separate systems use a dual port, omni-directional loop antenna which feeds two independent receivers that are controlled by two control panels.

Bearing information is displayed on the RDMIs. Selection of VOR on the EFIS control panel provides an EHSI dispaly of course, deviation, source(L/R VOR), and to/from information. A bearing angle relative to the transmitter is displayed on the RDMI and EHSI. Bearing, frequency, and course selection data are transferred in a digital format to the flight management computer system and to EFIS. Tuning can be done manually from the control panel, or it can be accomplished by commands from the flight management computer system. Audio goes directly to the interphone system.

The purpose of the weather radar system is to provide airborne weather detection, ranging and analysis. It can also be used to provide ground mapping as an aid to navigation by displaying significant ground objects and land contours.

The major components of this single system consist of an antenna, a transceiver and a control panel. The weather data is displayed on the two EHSIs. The weather radar system operates in the X-band with a peak power output of 125 Watts and a maximum range displayed of 320NM.

The received return signals are processed in the transceiver and the detected weather conditions and ground mapping information are sent to the EFIS symbol generators which control the displays on the left and right EHSIs. The EHSIs display green, yellow, and red for weather precitptation and magenta for turbulence.

The instrument landing system(ILS), see Figure 4-10-15, provides information showing deviation from the established glideslope and runway centerline(localizer). Three complete systems are installed, left, center, and right. Two dual antennas are provided for glideslope and two for localizer. One control panel has three channels, one for each system. Displays are shared with other navigation systems.

The marker beacon system is used to provide positive position checks enroute and during landing approaches. The maker beacon system consists of an antenna, a receiver module located in the left VOR receiver, and display lights located on the pilots' instrument panels.

Marker transmitters located in standard flight paths transmit a narrow vertical beam. As the aircraft flies over a given beam, the receiver detects the beam and lights the appropriate panel light.

4-10. Comm, Inst, and Nav. Systems 319

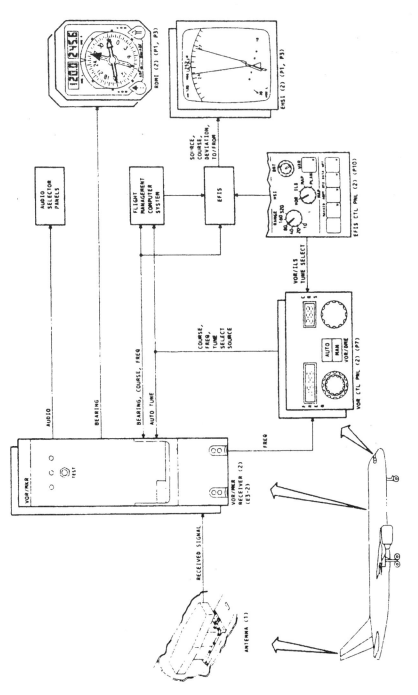

Figure 4-10-14 Boeing 757 VHF Omni range system(Boeing)

320 제4장 Aircraft System Description

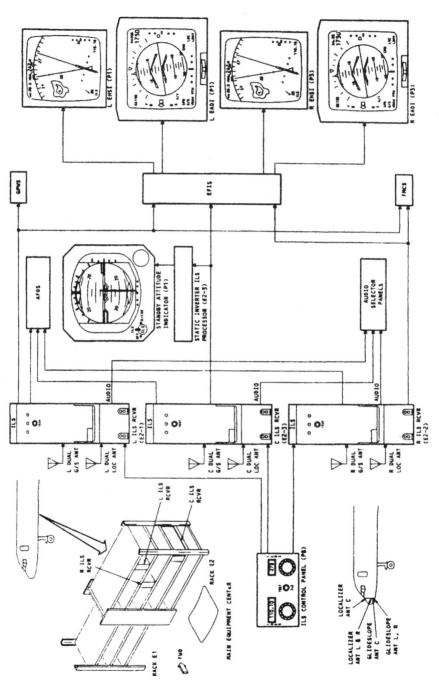

Figure 4-10-15 Boeing 757 ILS navigation system(Boeing)

During runway approaches, three different types of markers can be utlilized, with different pannel lights illuminating and different tones supplied through the audio selector panels to the pilots.

The radio altimeter(RA) system is designed to provide terrain clearance data(radio altitude) up to 2500ft, warning signals, and decision height information. This information is displayed primarily on the captain's and first offcer's electronic attitude director in dicators(EADI). Terrain clearance data is also used by the autopilot flight director system, ground proximity warning system, and EICAS.

The ground proximity warning system provides visual and aural warnings whenever the airplane is in danger of approaching terrain.

Below 2,500 feet, the system continuously monitors terrain clearance, descent rates, terrain closure rates, glide slope deviation, and flap/gear configuration so that alerts and warnings cna be generated whenever the airplane is in an unsafe condition due to terrain proximity.

E. Flight Instrument Systems

The flight instrument system(FIS), see Figure 4-10-16, provides airplane attitude, heading, vertical speed, DME distance, VOR and ADF bearing, flight director and ILS commands, flight management computer displays, and weather radar displays.

The flight instruments are located on both pilots' instrument panels. Each pilot's instruments consist of an electronic attitude director indicator(EADI), an electronic horizontal situation indicator(EHSI), a radio distance magnetic indicator(RDMI), and a vertical speed-indicator(VSI). The two electronic flight instrument control panels are located on left and right edges of the throttle quadrant stand.

The electronic flight instrument system(EFIS) operates as part of the flight management system to provide the EADIs with attitude and navigation information, and the EHSIs with ILS, VOR, or MAP displays both in a form suitable for accurate and rapid interpretation by both pilots. The EFIS also provides visual indications of failure warnings.

F. Flight Management Computer System

The autoflight system controls the airplane by operating the control surfaces in response to inputs from the pilot, primary navigation sensors, or steering commands from the flight management system.

322 제4장 Aircraft System Description

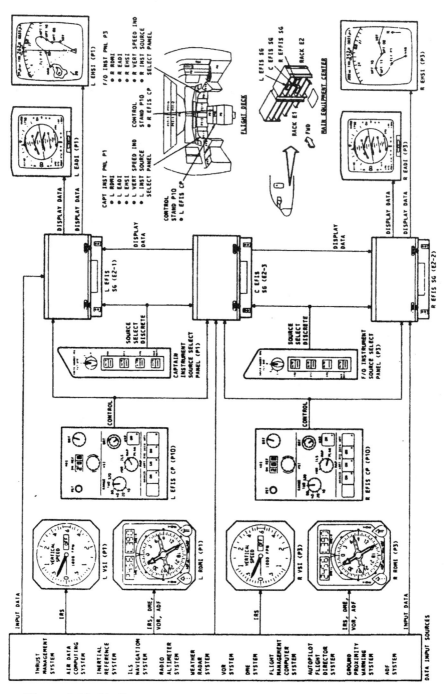

Figure 4-10-16 Boeing 757 EFIS flight instrument system(Boeing)

4-10. Comm, Inst, and Nav. Systems 323

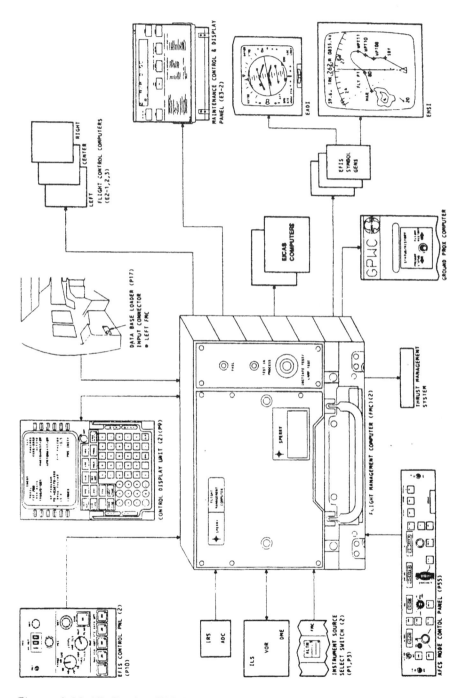

Figure 4-10-17 Boeing 757 flight management computer syystem (Boeing)

324 제4장 Aircraft System Description

The flight management computer system(FMCS), see Figure 4-10-17, uses inputs from navigation, engine, fuel system, and flight environment sensors along with stored and entered data to provide navigation performance and guidance control to the pilot and autopilot flight director system(AFDS).

The FMCS contains two flight management computers and two FMC control display units. The computers are located in the main euqipment center and control display units are located in the flight compartment.

Two flight management computers(FMC) receive inputs and provide outputs according to the mode selected on the FMC control display unit(CDU). The pilot may enter, retrieve or modify data or select modes from the FMC CDU. The flight management computer system may be used in two ways. The FMC may provide advisory data to the pilot enabling him to fly a selected course or profile, or the pilot may couple FMC inputs directly into the automatic flight director sysytem, causing the airplane to automatically follow a predetermined route. Armed and operational modes are displayed on the EADI with the autopilot, flight director or thrust management system engaged.

The flight management computer also provides both lateral and vertical navigation (LNAV and VNAV) based on a route specified by the flight crew and performance and navigation parameters stored within its memory. The crew monitors flight path parameters on the control display units(CDU) as well as on map and plan displays on the electronic horizontal situation indicators(EHSI). The CDU is also used to modify the route.

The autopilot flight director system(AFDS) provides pilot command and automatic modes for flight path control. Three flight control computers are installed. All functions for three axis control of the airplane are contained in each computer. Each computer drives a set of dedicated servos to move the flight control surfaces and provides input to the stabilizer/elevator asymmetry module(SAM) for automatic stabilizer trim.

G. Yaw Damper System

The yaw damper modules use inputs from the air data computers and the inertial reference computers to derive rudder commands appropriate to flight conditions. These commands go to the yaw damper servos. Additonally, the modules monitor system operation and provide both manually initiated and automatic system testing.

The yaw damper servos use electrical commands from the yaw damper modules to control hydraulic flow to an actuator pistion. This motion is linked in series with manual and autopilot inputs to the rudder power control actuators.

H. Thrust Management System

The thrust management system performs thrust limit calculations and controls the throttles for full flight regime autothrottle operation. One thrust management computer moves the throttles and drives the displays on the EICAS display. Thrust limits are calculated based on selected mode and operating conditions.

Autothrottle functions control thrust, mach/airspeed, rate of altitude change or throttle retard rate as selected. A fast-slow command indicator reflects speed error in mach/airspeed mode. Limit functions operate in all modes to prevent overboost or overspeed.

I. Digital Flight Data Recorder System

The digital flight data recorder system(DFDRS) provides the capability of recording the most recent 25 hours of flight parameters on magnetic tape, housed in a crash proof container.

The flight parameters that must be recorded as required by the FAA are: GMT-time, pressure altitude, computer airspeed, acceleration, compass heading, pitch and roll altitudes, control surface positions, engine status and performance, and VHF and HF radio transmissions. An underwater locator beacon is installed on the front face of the DFDR.

4-11. Miscellaneous Aircraft Systems and Maintenance Information

1) Portable Water Systems

A. Lockheed L-1011 Potable Water System

The portable water system, shown in Figure 4-11-1, distributes drinkable water to lavatory sinks, drinking fountain and coffee makers in the galleys. Chlorinated portable water is stored in a fiberglass 150 gallon tank located in the lower galley sidewall area. Tank quantity, which varies between airline, is determined by the length of a standpipe in the overflow line, as shown in Figure 4-11-2.

The air space above the water level is pressurized through the air supply port by two on-board compressors, illustrated in Figure 4-11-3, or by a ground air source connected at the service panel. Compressor output passes through position indication check valves, which monitor the output pressure and cause annunciatiors at the flight station to illuminate if a 5 PSI differential is sensed. Air then enters the tank through a screen filter, relief valve, and check valve. The filter contains a red indicator which is displayed when the filter is 80% clogged.

Compressor control switches monitor pressure in the tank to turn on the associated compressor when tank pressure is less than 30 PSI, and turn off the compressor when tank prssure reaches 35 PSI. The air is directed to the tank through a 35 PSI regulator and a check valve.

If both compressors are inoperative, water can be made available to the user system by half filling the tank with water, and pressurizing the remaining air space from a ground source. On demand, water under pressure exits the tank through an outlet at the lower center which is plumbed to all using facilities.

The material composition of most distribution lines is corrosion-resistant steel; however, titanium, monel and flexible Teflon® hoses are used in some areas.

To prevent freezing, some distribution lines are heated by wrap-around heater blankets; heater jackets protect the fill/overflow valve and the drain valve. Two manually-operated shutoff valves are located in the distribution lines to permit isolation of galley facilities for maintenance purposes.

A 3/4" fill connection at the water service panel provides for attachment of a water supply line from ground servicing equipment. Water enters the tank through the open fill╱

4-11. Misc. A/C Systems and Maint' Information 327

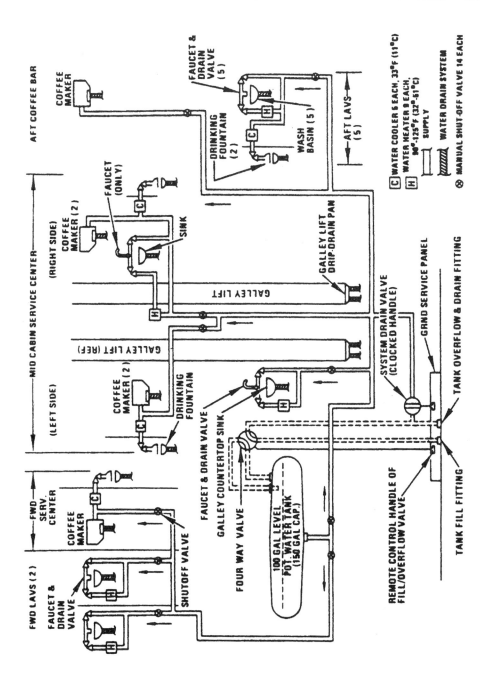

Figure 4-11-1 L-1011 Potable water distribution diagram (Lockheed)

328 제4장 Aircraft System Description

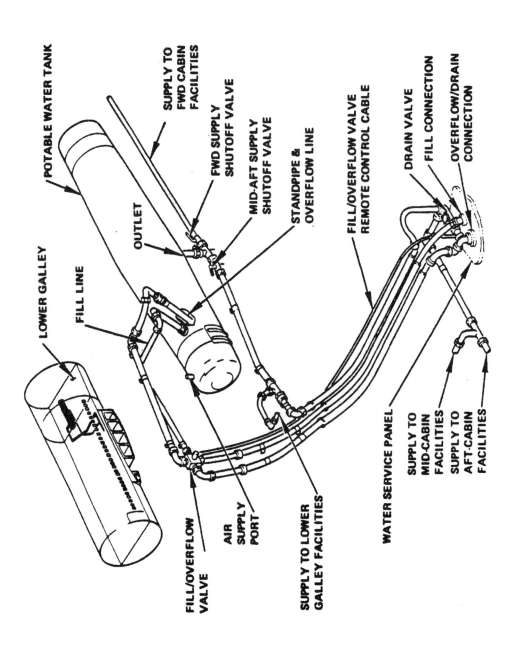

Figure 4-11-2 L-1011 Potable water storage system(Lockheed)

4-11. Misc. A/C Systems and Maint' Information 329

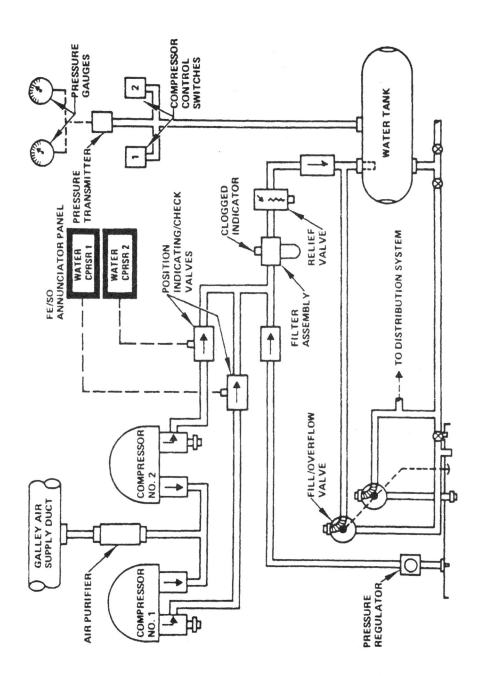

Figure 4-11-3 L-1011 Water pressure system (Boeing)

overflow valve, which can be operated remotely at the service panel through a remote control cable, or directly at the valve by the valve handle. When the tank water level reaches the standpipe, overflow water returns to the service cart via the overflow line. When closed, the fill/overflow valve enables pressurization by interconnecting the fill port/line with the standpipe/overflow line.

The normally closed drain valve can be opened by operating its handle at the service panel. When the drain valve and supply shutoff valves are open, the entire system can be drained to a holding cart attached to the overflow/drain connection. The system must be drained if the aircraft is going to be parked without electrical power in below freezing weather for any length of time.

2) Waste Systems

A. L-1011 Waste System

The waste system, Figure 4-11-4, consists of independent forward and aft subsystems which are functionally the same and physically similar. Both installations are self-contained, recirculating liquid chemical systems which process and store lavatory toilet wasts products. Forward lavatory toilets are served by a 40-gallon waste tank; aft lavatory toilets, by an 80-gallon tank.

Forward and aft service panels contain provisions for draining, flushing and pre-charging the associated tank. The service panels also contain indicating lights for tank liquid level indication.

Each tank is pre-charged prior to flight with a dye-deodorant-disinfectant chemical flushing liquid; the forward tank with 15 gallons, the aft with 20 gallons.

In each system, pressure for toilet flushing is supplied by three tank-mounted motor/pump/filter assemblies, which operate one at a time in selective rotational sequence. One pump develops sufficient pressure to flush all subsystem toilets simultaneously. The pumps are controlled by solid-state circuitry in a logic control box mounted near the tank.

A flush valve on each toilet in the system is also controlled by the logic box. Flush signals are initiated at each toilet by a flush control lever/switch assembly. The logic control boxes are identical; each contains line-replaceable printed circuit cards, including a separate 12-second flush timer card for each toilet in the subsystem and space for spare timer card storage. Each logic box also contains externally-accessible switches and indicator lights for testing motor/pump/filter assemblies.

4-11. Misc. A/C Systems and Maint' Information   331

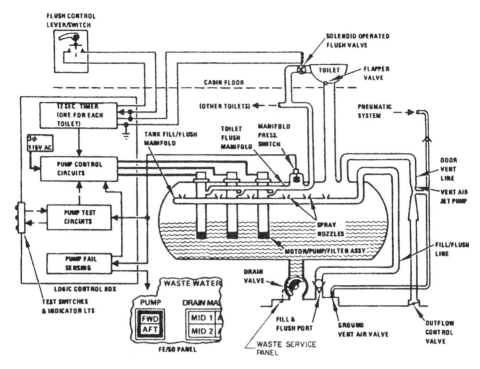

Figure 4-11-4 L-1011 Waste disposal functional diagram(Lockheed)

The waste panel, shown in Figure 4-11-5, at the flight engineer station, contains forward and aft pump dial lights for annunciation of pump failure in the associated subsystem.

A flush signal is generated when a flush control lever is momentarily pressed. Logic box circuitry opens the associated flush valve and activates the next pump in rotational sequence. At the end of 12 seconds the flush valve is closed and the pump is shut off, unless its operating time is extended through activation of other timers by concurrent flush signals.

Wasts tank odors are ducted overboard through a vent line which terminates near the service panel. Air pressure for tank venting can be supplied from one of three sources: cabin pressure, the pneumatic system, or ground air. When the aircraft is pressurized, cabin air passes through the toilets into the waste tank and exhausts overboard through the vent line and outflow control valve. If the cabin is not pressurized, odors are drawn from the tank by a vent air jet pump in the vent line, which can be operated from air pressure supplied by the pneumatic system, or from a ground source through the ground vent air valve at the service panel.

332 제4장 Aircraft System Description

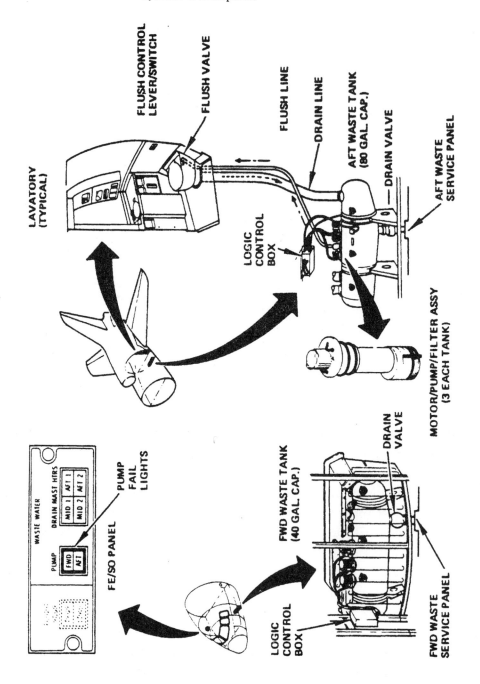

Figure 4-11-5　L-1011 FWD and AFT Waste system(Lockheed)

4-11. Misc. A/C Systems and Maint' Information   333

The waste tank is drained, flushed and filled at the waste service panel(Figure 4-11-4). It is gravity drained to a suitable ground holding cart through a manually-operated drain valve, and is pressure flushed/filled through a self-sealing fill/flush port. Fill or flush liquid enters the tank through a fill/flush manifold which contains six spary nozzles. The nozzles direct flush liquid spray to assure complet cleaning action.

With the drain valve closed, the tank is pre-charged through the fill/flush port and spray nozzles with a dye-deodorant-disinfectant liquid chemical. The service panel also contains indicating lights for determining the quantity of accumulated waste in the tank.

B. 747-400 Toilet Waste System

The 747-400 toilet waste system uses a vacuum systm to collect, transport and store toilet waste. The waste system, shown in Figure 4-11-6, provides toilets throughout the main deck and upper deck passenger cabin.

The system uses portable water to flush the toilet bowls, and vacuum(differential pressure) to transport the waste through toilet drain lines to waste tanks in the bulk cargo compartment. the toilet flushing cycle is controlled by a flush control unit. This unit

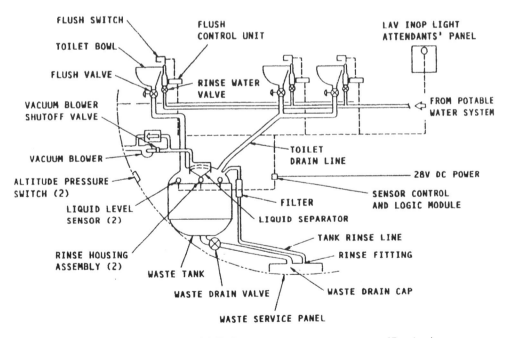

Figure 4-11-6 747-400 Toilet waste system summary(Boeing)

334 제4장 Aircraft System Description

sequences and times the cycle including the rinse water valve, flush valve and vacuum blower operation.

The vacuum(differential) is created by a vacuum blower at lower altitudes and differential pressure at high altitudes. At altitudes below 16,000 feet(12,000 feet on descent) the vacuum blower is not operating, and the blower shutoff valve is closed. At high altitudes, the differential pressure between the cabin and ambient provides the vacuum.

All air leaving the waste tanks passes through a liquid separator, Waste tank level is monitored by two level sensors in each waste tank. The sensors are connected to a sensor and logic control module.

When a waste tank reaches capacity, the level monitoring system prevents the associated toilets from flushing, and illuminates a LAV INOP light on the attendant's panel. all waste tanks are serviced from one waste service panel shown in Figure 4-11-7.

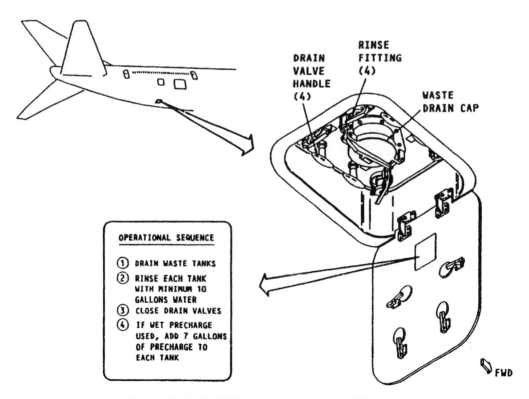

Figure 4-11-7 757 Waste service panel(Boeing)

4-11. Misc. A/C Systems and Maint' Information

The panel contains a waste drain cap and rinse fittings. The drain cap is opened to provided a connection for ground service equipment. When the drain valves are opened, waste tank contents flow out this connection.

After draining, the tanks are rinsed with flushing liquid through the rinse fitting on the service panel. The liquid passes through a filter before entering the rinse housing assemblies in the tank. After rinsing, a wet pre-charge may be added to the tanks.

3) Lighting Systems

A. Boeing 757 Aircraft Lighting Systems

Aircraft lighting systems provide illumination of several different areas of the aircraft. Some areas that require lights are the flight, passenger, cargo, and service compartments. In addition to compartment lights, exterior and emergency lighting systems also provide light in critical areas, in and around the aircraft.

The exterior lights provide illumination of the ground during landing and taxi operations. The landing lights can also be used to make the aircraft more visible to air traffic controllers and other aircraft in flight.

The 757 has four landing lights(see Figure 4-11-8) which are installed with two in the wing roots and two on the nose landing gear.

The wing root lights shine horizontally and the nose gear lights are aimed downward on a typical glide slope angle.

Other exterior lights include the wing illumination lights which are mounted on each side of the fuselage to light up the leading edge of the wing and the engine necelles.

The runway turnoff lights are mounted on the nose landing gear and illuminate the area to either side of the aircraft.

Anti-collision lights(strobe lights) are mounted on the top and bottom of the fuselage and on each wing up. The fuselage lights are covered with a red lens and the wing lights are covered with a clear lens.

There are two position lights mountd on each wing tip facing forward and aft as shown in Figure 4-11-8. The aft facing lights are covered with a clear lens while the forward facing lignts have a red lens on the left wing and a green lens on the right wing.

Two quartz halogen logo lights are positiond on the top surface of each horizontal stabilizer to illuminate the vertical stabilizer.

336 제4장 Aircraft System Description

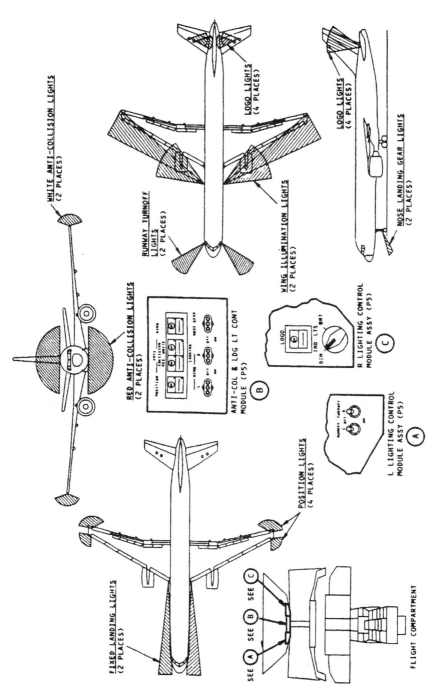

figure 4-11-8 757 Exterior lights(Boeing)

4-11. Misc. A/C Systems and Maint' Information

B. Flight Compartment Lighting

The main instrument panel floodlights are controlled from switches on the P7 lighting panels, see Figure 4-11-9. These lights are fluorescent tubes and vary the length of time they are switched on to provide variable intensity.

The glareshield and aisle stand floodlights are controlled from rheostats located on the P5 overhead panel. There are three dome lights in the 757 which are controlled by a rheostat. Lighting for all the instrument and circuit breaker panels is controlled from the lighting control panels. The bulbs used for faceplate illumination are soldered on a circuit strip attached to the rear of the faceplate.

Map lights and chart lights controlled from the P7 panels illuminate the pilot's letdown chart holders and lap area. The flight compartment also includes utility(kit) lights which are moveable, two colored, spot lights for miscellaneous use. The override light switch shown on the P5 panel in Figure 4-11-9 can be used as a means of turning on all of the instrument panel floodlights and dome lights from one switch.

C. Passenger Compartment Lights

General illumination of the passenger cabin is with dual intensity fluorescent ceiling and sidewall lights operated with switches on the forward attendant's panel to the bright or dim mode, as shown in Figure 4-11-10. Incandescent lighting is used for additional spot lighting of work and access areas. A passenger reading light and switch is provided for each passenger seat.

A FASTEN SEAT BELT and a NO SMOKING sign is installed in each passenger service unit. Lavatory signs are located at each lavatory location and an occupied sign is installed as part of each lavatory sign to indicate when lavatories are in use. Exit signs mark each exit and also support the passenger, lavatory and crew call lights.

Laboratories have fluorescent mirror lights for general illumination and incandescent dome lights for dim illumination. Inflight, the mirror lights are turned on and off by operation of the lavatory door latch switches. They illuminate when the door is closed and locked and go out when unlocked.

D. Cargo Compartment Lights

Forward and aft cargo compartment lighting is provided for the interior and exterior of each compartment. The forward cargo compartment lights are powered by 115 VAC and are enabled when the cargo door is not fully closed. The lower hinge proximity switch

338 제4장 Aircraft System Description

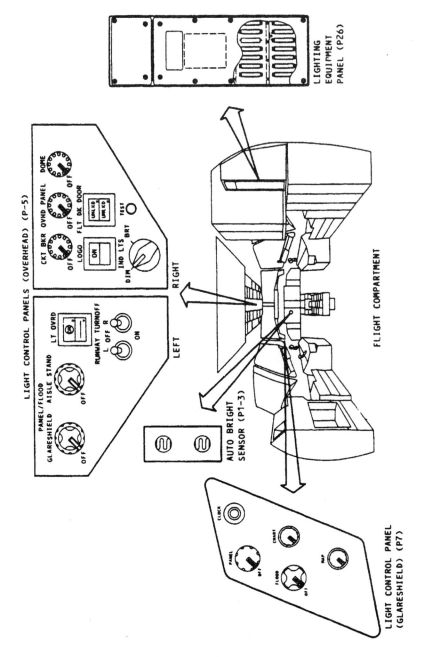

figure 4-11-9 757 Flight compartment lights(Boeing)

4-11. Misc. A/C Systems and Maint' Information 339

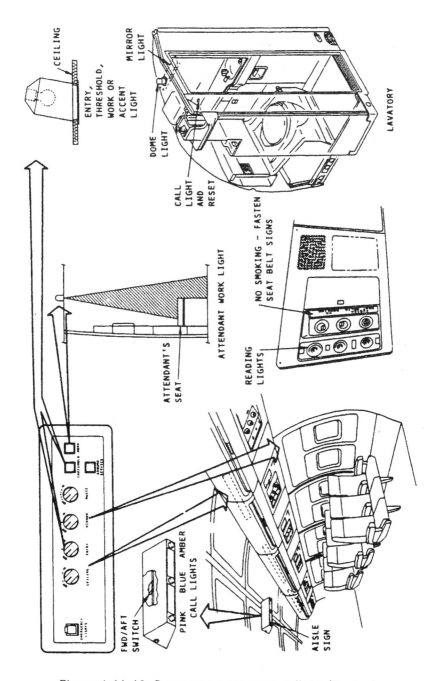

Figure 4-11-10 Passenger compartment lights (Boeing)

provieds a ground which energizes the forward cargo compartment lights. The forward cargo compartment lights switch controls the five lights in the ceiling and the five lights in the door. The light switch is located just aft of the door. The aft cargo compartment lights switch controls the eight lights in the ceiling and the five lights in the door.

E. Service Compartment Lights.

Service compartment lighting is provided in the main and nose wheel wells and in the E/E equipment compartments. Service compartment lighting is also provided in the air conditioning. APU, and tail cone compartments. Service lighting generally used to inspect or perform work in these service compartments. Cargo and service compartment lighting is shown in Figure 4-11-11.

4) Emergency Equipment

A. Boeing 757 Emergency Lights

The emergency lights provide illumination of the passenger cabin and escape slide routes, using battery packs as the power source. The lights can be turned on manually or set to come on automatically in the event of failure of the primary lighting power.

The lights, illustrated in Figure 4-11-12, consist of EXIT sign modules over each door, exit indicators near the floor at each exit and over the main aisle between doors, area lights on ceiling and floor, in the cross aisles between doors, main aisle lights evenly spaced along the main aisle, floor mounted lights at 20" intervals on the left hand side of the aisle and slide lights externally mounted aft of each door and directed to illuminate the slide path.

Ten battery packs mounted in the passenger cabin ceiling area are used as the power source for these lights. These battery packs are mounted in pairs in the area of each door. Each pair of packs provides emergency power for the illumination of the adjacent lights.

The primary operational control for the emergency lights is a three position guarded switch on the pilot's overhead panel. The switch has an On, an OFF, and a guarded ARMED position. Whenever this switch is in other than the ARMED position, the amber UNARMED caution light will illuminate, and an EMER LIGHTS alert message will be sent to the EICAS system. If aircraft power is lost and the emergency light switch is in the ARMED position, the emergency lights will be illuminated automatically, powered by the battery packs described earlier. The normal emergency light switch position when passengers are onboard the aircraft is the armed postion.

4-11. Misc. A/C Systems and Maint' Information 341

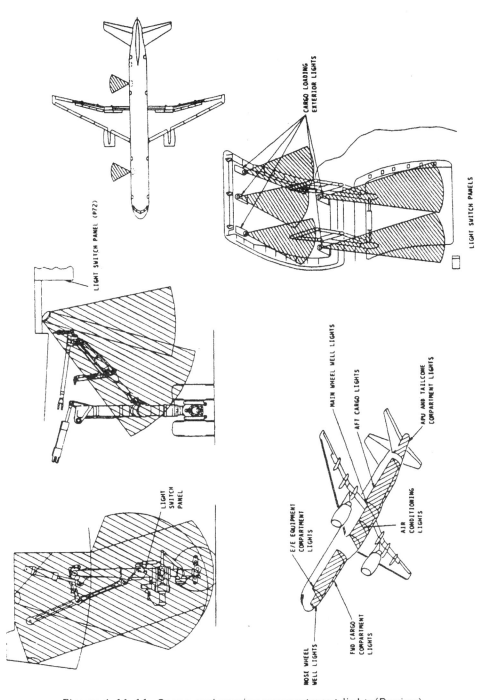

Figure 4-11-11 Cargo and service compartment lights (Boeing)

342 제4장 Aircraft System Description

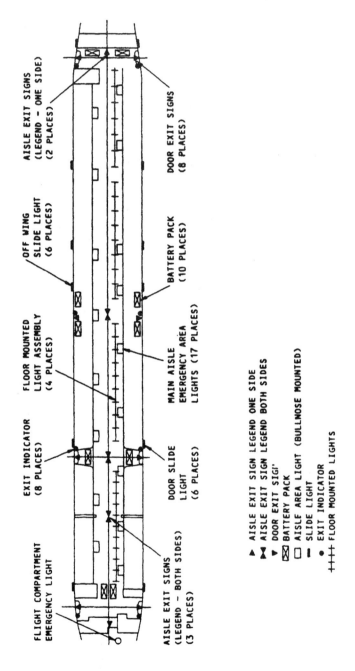

Figure 4-11-12 757 Emergency lights layout(Boeing)

4-11. Misc. A/C Systems and Maint' Information   343

B. Boeing 737-300 Emergency Equipment

Clearly marked exit lights are located throughout the passenger cabin to indicate the approved emergency exit routes. All of the lights are powered by individual nickel cadmium batteries with a charging monitoring and voltage regulator circuit.

Floor proximity emergency escape path lighting consists of locator lights spaced at approzimately 40-inch intervals down the left-hand side of the aisle. Lighted arrows point to overwing exits and a light EXIT indicator is near the floor by each door and overwing exit. Escape path markings are provided for visual guidance for emergency cabin evacuation when all sources of cabine lighting more than four feet above the aisle floor are totally obscured by smoke.

Exterior emergency lights illuminate the escap slides, as shown in Figure 4-11-13. The fuselage installed escape slide lights are adjacent to the forward and aft service and entry doors. Two lights are also installed on the fuselage to illuminate the overwing escape routes and ground contact area.

Emergency evacuation may be accomplished through four entry/service doors and two overwing escape routes and ground contact area.

Cockpit crewmembers may evacuate the airplane through two sliding cockpit windsows.

The emergency evacuation routes are shown in Figure 4-11-14. Two escape hatches are located in the passenger cabin over the wings. These are plug-type hatches and are held in place by mechanical locks and airplane cabin pressure. The hatches can be opened from the inside or from the outside of the airplane by a spring-loaded handle at the top of the hatch.

C. Escape Slides.

When an emergency dictates rapid evacuation of the airplane on the ground, it may be necessary to activate and use the emergency escape slides shown in Figure 4-11-14. The slides are inflatable rubber/nylon units which are stored in compartments on the bottom inner face of the forward entry, aft entry, and galley service doors.

The slide incorporates a retainer(girt) bar, illustrated in Figure 4-11-15, which is normally stowed in special stowage clips on the compartment cover. Before texiing, this bar is removed from the hooks and fastened to brackets located on the floor of the airplane. It remains there throughout the flight.

If an emergency evacuation is needed upon landing and the door is opened, tension on the girt bar will cause the compartment latch to separate, allowing the compartment to open

344 제4장 Aircraft System Description

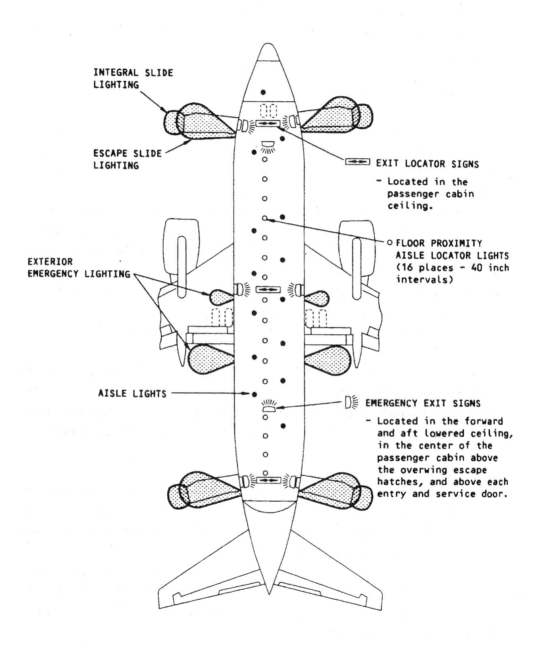

Figure 4-11-13 737 Exterior emergency exit light location(Boeing)

4-11. Misc. A/C Systems and Maint' Information 345

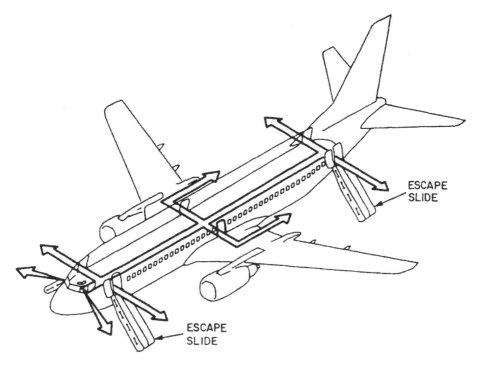

Figure 4-11-14 737 Emergency evacuation routes (Boeing)

and the slide to automatically deploy outboard of the door opening.

For normal door operation, the girt bar is stowed in the spring loaded retaining clips at the bottom of the door. These clips require the girt bar to be snapped into position and not just laid on top of the clips. Improper stowage of the girt bar could result in inadvertent deployment of the slide. The slide cover has a viewing window for checking the pressure in the escape slide inflation bottle.

Inflation will begin during slide deployment. Automatic inflation requires approximately 5 seconds.

Should the automatic inflation system fail, the manual inflation handle must be pulled completely clear of the slide to effect proper inflation. The manual inflation handle is labeled PULL and is visible when the slide is ejected from its container.

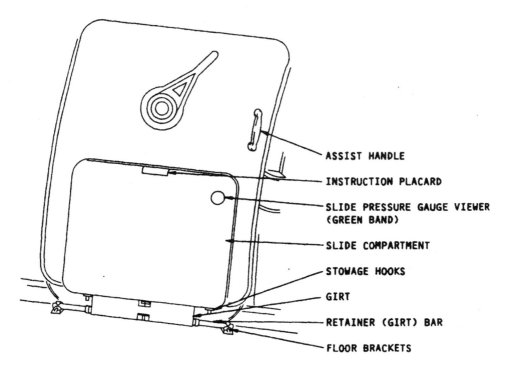

Figure 4-11-15 Escape slide compartment and girt bar(Boeing)

5) Equipment Cooling Systems

A. Boeing 737-300 Cooling System

EFIS equipment, circuit breaker panels in the cockpit, and electronic equipment in the E & E compartment are cooled by the equipment cooling system shown in Figure 4-11-16. Warm air, caused by the heat generated by the aircraft's electrical components, is ducted away by the selected AC powered fan.

On the ground, or with the cabin differential pressure less than 2.5 PSI, the exhaust fan air is blown through a flow control valve and exhausted out the bottom of the airplane.

With increasing airflow at great cabin differential pressures, the flow control valve closes. This action will occur when the aircraft is pressurized at altitude and has a cabin pressure differential of more than 2.5 PSI. At this point, the cabin air to outside air pressure differential is sufficient to supply the equipment cooling system. Warm air from the

4-11. Misc. A/C Systems and Maint' Information 347

electronic equipment cooling system is then diffused around the forward cargo compartment and out the forward outflow valve. This insures adequate heating of the forward cargo at high(colder) altitudes.

The no-airflow detectors in the ducting just forward of the equipment cooling fans consists of a thermal switch. Loss of airflow due to failure of an equipment cooling fan causes increasing heat to activate the thermal switch, illuminating the respective equipment cooling OFF light. Selecting the alternate fan should restore airflow passing the switch, and extinguish the OFF light.

Additional thermal switches are located in the E&E compartment. If an overtemperature occurs on the ground, alerting is also provided through the ground crew call horn in the nose wheel well.

B. L-1011 Instrument and Electronics Cooling System

The instrument cooling system, shown in Figure 4-11-17, uses a fan which runs all the time the aircraft is using AC power. The fan draws air from below the flight engineer panel and supplies it to three nozzles behind the flight engineer panel, to one nozzle behind the engine instrument panel, and to two manifolds behind the pilots' instrument panels. From the manifolds, the air is directed at the cases of the horizontal situation indicator(HSI) and the attitude director indicator(ADI), and to the surrounding areas for cooling. Behind the flight engineer panel, the air is generally directed to all instruments.

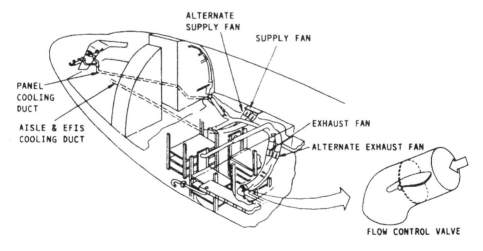

Figure 4-11-16 737 Equipment cooling components diagram(Boeing)

348 제4장 Aircraft System Description

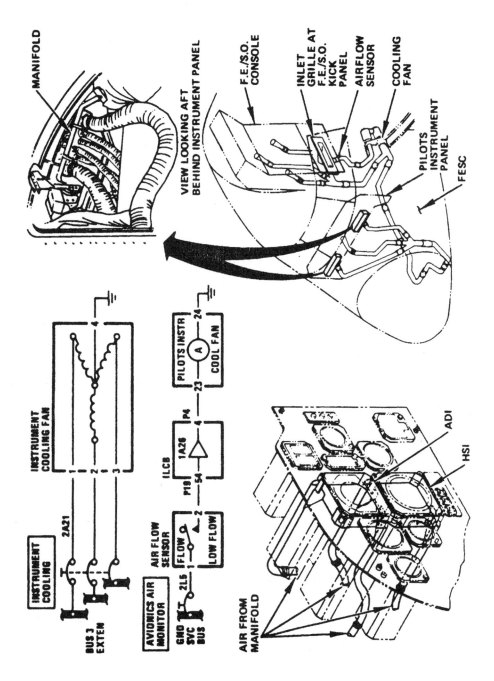

Figure 4-11-17 L-1011 Instrument cooling system(Lockheed)

4-11. Misc. A/C Systems and Maint' Information 349

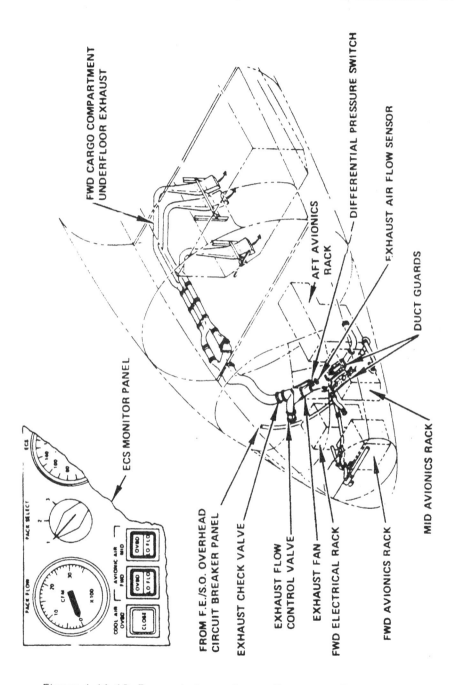

Figure 4-11-18 Forward electronics cooling system(Lockheed)

When insurfficient airflow is available for adeuate instrument cooling, an airflow sensor, mounted on the fan supply duct behind the flight engineer panel, will provide a signal to illuminate the PILOTS INSTR COOL FAN light on the flight engineer annunciator panel.

A filter is installed in front of the grill to keep dirt and debris in the flight station from being drawn through the fan. The filter is held in place with Velcro® tape, so that when the filter becomes dirty, it is easy to remove and wash.

The forward electronics system shown in Figure 4-11-18, uses a fan to draw air through the equipment and passes it under the floor of the forward cargo compartment where it vents overboard through the forward pressurization outflow valve. When the aircraft is on the ground, the exhaust flow control valve will open, even though the fan is still running. This exhausts the cooling air directly overboard.

C. 757 Equipment Cooling System

The Boeing 757 equipment cooling system is divided into a forward and an aft system. The forward system, illustrated in figure 11-19, cools the equipment by using blow through and draw through air cooling. The forward system also incorporates an air cleaner, an overheat, a low flow, and a smoke detection system.

The low flow sensor installed in a branch of the exhaust ducting monitors airflow in the system and a smoke detector monitors the exhaust air for the presence of smoke. The overboard exhaust air for the presence of smoke. The overboard exhaust valve, connected to the ducting downstream from the left recirculation fan, seves as a back-up source for forward exhaust system airflow during flight and allows a major portion of the forward exhaust air to be discharged overboard during ground operation.

The aft equipment cooling system uses only draw through air to provide cooling for the equipment in the electrical racks. There is no smoke or low flow detection installed in the aft system. Both systems normally operate automatically through the use of relays.

The EICAS computers monitor the equipment cooling system for proper operation and provide messages to indicate malfunctions or normal operation. Indications will also be provided on the equipment cooling control panel, the air conditioning control panel and through the ground crew call system.

4-11. Misc. A/C Systems and Maint' Information 351

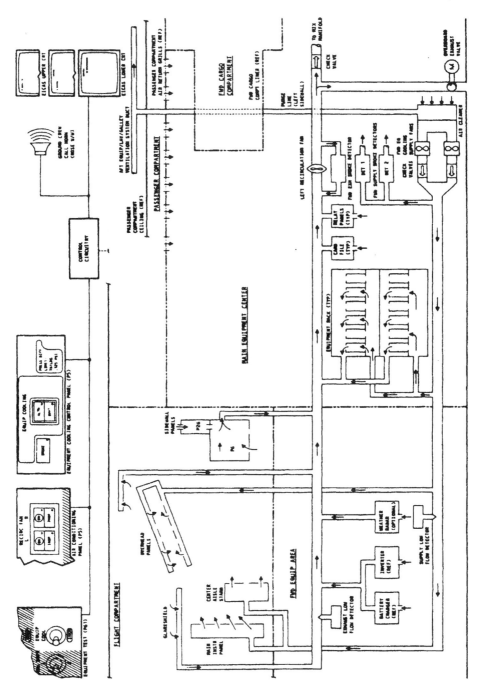

Figure 4-11-19 757 Forward equipment cooling system (Boeing)

Abbreviation

— A —

A/C	Aircraft ; 항공기
AC	Alternating Current ; 교류(전류의)
ACL	Allowable Cabin Load ; 항공기의 화물과 승객 탑재 가능량
AD	Airworthiness Directive ; 감항성 개선 명령
ADF	Automatic Direction Finder ; 자동 방향 탐지기
ADI	Anti-Detonation Injection ; 실린더 내의 이상 연소를 억제하는 연료 분사를 말함
ADI	Attitude Direction Indicator ; 항공기의 자세 전체를 한눈에 알아볼 수 있게 한 계기
AF	Air Force ; 공군
AIP	Aeronautical Information Publication ; 항공로 지도
AN Spec.	Airforce-Navy Aeronautical Specification ; 미국 공·해군 항공 규격
AN Standard	Airforce-Navy Aeronautical Standard ; 미국 공·해군 항공 표준
AOG	Aircraft On the Ground ; 부품이 없어 정비를 중단하고 있는 항공기를 말하며 부품의 긴급 조달 용어이다.
A/P	Autopilot ; 자동 조종 장치
APU	Auxiliary Power Unit ; 보조 동력 장치
ARB	Air Registration Board ; 항공국(영국)

ARINC	Aeronautical Radio Inc. ; 에어링크 에어링크의 활동은 국내 및 해외의 항공 지상 무선국의 작동, 지상과 하늘의 적합성을 달성하기 위한 시스템의 각종 요구의 실천, 이들 요구에 맞는 전파를 할당하여 기술 정보의 교환 등으로 성립되고 있다.
ARSR	Air Route Surveillance Radar ; 항공로 감시 레이더
ARTS	Automated Radar Terminal System ; 터미널 관제 정보 처리 시스템으로 이차 감시 레이더(SSR)에 의해 얻어지는 항공기의 운항에 관한 정보를 전자 계산기로 처리해서 항공기의 자동 식별을 하는 것 외에 레이더 표시면상에 항공기의 편명, 비행 고도, 대지 속도 등의 비행 정보를 표시한다.
ASDE	Airport Surface Detection Equipment ; 공항 내 지상 감시 레이더를 말하며 공항 지표면의 교통을 감시하고 항공기나 차량의 움직임을 관제하여 그들의 안전을 확보하기 위한 것이다.
ASR	Airport Surveillance Radar ; 공항 감시 레이더로서 공항 주변 공역에 있는 항공기의 위치를 탐지하여 착륙, 이륙시에 항공 교통 관제에 사용된다.
ATA	Actual Time of Arrival ; 실제의 도착 시각
ATA	Air Transport Association ; 항공운송협회
ATC	Air Traffic Control ; 항공 교통 관제
ATD	Actual Time of Departure 실제의 출발 시각
ATIS	Automatic Terminal Information Service ; 비행장 정보 방송 업무를 말하며 기상 정보, 비행장의 상태, 항공 보안 시설의 운용 상황 등의 정보를 항상 자동적으로 항공기에게 방송하는 업무이다.
ATK	Available Ton Kilometer ; 유효 톤 킬로미터, 각 비행 구간의 유효 중량에 그 구간의 거리를 곱한 것

AWP	Awaited Parts ; 부품이 없어 수리를 중단하고 있는 장비품을 말하며 부품의 긴급 조달 용어이다.

— B —

BFE	Buyer's Furnishing Equipment ; 항공기를 구입할 경우. 그 항공기에 장착하는 어떤 특정한 계기에 대해서 항공기 제조 회사가 준비하지 않고 구매자가 준비하여 제조 회사에 장착을 의뢰하는 것이 있다. 이들 계기를 말한다.
BLC	Boundary Layer Control ; 경계층 제어
BMEP	Brake Mean Effective Pressure ; 제동 평균 유효 압력

— C —

CAR	Civil Aviation Regulations ; 항공법 시행규칙
CAS	Calibrated Air Speed ; 교정 대기 속도. 대기 속도계 계통의 오차 수정을 한 대기 속도계의 눈금이 나타내는 속도를 말한다.
CC	Cubic Centimeter ; 입방 센티미터
CB	Circuit Breaker ; 서킷 브레이커
CCW	Counterclockwise ; 반시계 방향
CFE	Customer's Furnishing Equipment ; BFE와 같음
CG	Center of Gravity ; 중력 중심
C'K	Check ; 점검
CW	Clockwise ; 시계 방향

— D —

DC	Direct Current ; 직류(전류의)
DME	Distance Measuring Equipment ; 거리 측정 장치
DMET	Distance Measuring Equipment TACAN
DN	Down ; 아래로

— E —

EAS	Equivalent Air Speed ; 등가 대기 속도, 대기 속도계 계통의 오차 및 특정한 고도에서의 단열 압축 흐름의 효과에 대해 수정을 한 대기 속도계의 눈금이 나타내는 속도를 말한다.
EHP	Equivalent Horse Power ; 상당 마력
EPNdB	Equivalent Perceived Noise dB ; 실효 감각 소음 레벨
EPR	Engine Pressure Ratio ; 엔진 압력비
ESHP	Equivalent Shaft Horse Power ; 상당 축마력
ETA	Estimated Time of Arrival ; 예정 도착 시각
ETD	Estimated Time of Departure ; 예정 출발 시각

— F —

FAA	Federal Aviation Administration ; 미국연방항공국
FAR	Federal Aviation Regulation ; 미국연방항공규칙
FCC	Federal Communications Commission
FCU	Fuel Control Unit ; 연료 조절 장치

F/E	Flight Engineer ; 항공 기관사
F/F	Fuel flow ; 연료 유량
FH	Flight Hour ; 비행 시간
FIR	Flight Information Region ; 비행 정보구로서, 각국이 항공 교통 업무를 담당하는 구역을 나타내며 ICAO에서 결정된다.
FL	Flight Level ; 비행 고도
F/O	First Officer ; 부조종사
FOD	Foreign Object Damage ; 외부 이물질에 의한 손상

— G —

GCA	Ground Controlled Approach PAR ; 정밀 진입 레이더를 사용해서 계기 비행 방식에 의해 비행하는 항공기에 대해 착륙의 유도를 하는 착륙 유도 관제 업무를 말한다.
GPU	Ground Power Unit ; 지상 전원 장치
GSE	Ground Support Equipment ; 지상 지원 장비

— H —

HF	High Frequency ; 단파
HP	Horse Power ; 마력

— I —

IAS	Indicated Air Speed ; 지시 대기 속도, 대기 속도계 계통의 오차, 즉 위치 오차 수정을 하지 않은 항공기에 부착

	된 피토 정압식 대기 속도의 눈금이 나타내는 속도를 말한다.
IATA	International Air Transport Association ; 국제 항공 운송 협회의 약칭으로 세계의 정기 항공회사의 단체로 1945년 4월에 하바나에서 결성되었다. 본부는 스위스의 제네바에 있고 각지에 지부가 있다. 안전하고 경제적인 항공 수송을 발달시켜 항공에 의한 무역을 추진하며, 이에 관한 여러가지 문제를 연구하는 업무간의 협력 기관으로서 ICAO 기타의 국제 단체와 협력하는 것이 목적이다.
IACO	International Civil Aviation Organization ; 국제 연합의 전문 기관의 하나인 국제 민간 항공 기관의 약칭. 1944년의 국제민간항공조약(통상 시카고조약)을 바탕으로 하여 설립된 기관으로 국제 민간 항공의 안전하고 질서 있는 발달 및 국제 항공 운송 업무의 건전하고 적정한 운영을 도모하는 것을 사명으로 하여 각종 활동을 하고 있다.
ICBM	Intercontinental Ballistic Missile ; 대륙간 탄도탄
ID	Identification ; 식별
IFR	Instrument Flight Rules ; 계기 비행 방식
ILS	Instrument Landing System ; 계기 착륙 장치
INS	Inertial Navigation System ; 관성 항법 장치를 말하며 항공기의 가속도를 적분 계산해서 속도와 거리를 얻고 항공기의 위치, 목적지까지의 거리, 비행 시간 등 항법상 필요한 자료를 얻는 자려 항법 장치이다.
IRAN	Inspection and Repair As Necessary ; 아일런 작업
IRBM	Intermediate Range Ballistic Missile ; 중거리 유도탄

— J —

JATO	Jet Assisted Takeoff ; 이륙시에만 사용하는 보조적인 젯트 추진 장치

— K —

KCAB	Korea Civil Aviation Bureau ; 교통부 항공국
kHz	Kilohertz ; 킬로헤르쯔(주파수의)
kt	Knot ; 노트

— L —

LE or L/E	Leading Edge ; 리딩에이지
L/F	Load Factor ; 여객 화물의 탑재 가능량에 대해 실제로 탑재한 여객, 화물의 비를 말함.
LG or L/G	Landing Gear ; 착륙 장치
LH	Left Hand ; 좌측
LORAN	Long Range Navigation ; 로란 장치

— M —

M	Mach ; 마하, 음속을 1로 하여 나타낸 속도의 단위
MH	Man Hour ; 시간당 작업량
MIL	Military Specification ; 미국 육군 규격
MLG	Main Landing Gear ; 메인 랜딩기어
MM	Maintenance Manual ; 정비 교범, 정비 기준

mph	Mile Per Hour ; 속도의 단위로 1시간당의 마일 수를 나타낸다.
MRB	Material Review Board, Maintenance Review Board

— N —

NASA	National Aeronautics and Space Administration ; 미국항공우주국
NDB	Non-directional Radio Beacon ; 무지향성 무선 표식 시설을 말하며 항공로를 따른 요소 또는 공항에 설치되어 있다. 지상국에서 무지향성의 전파를 발사하여 항공기에서 방향 탐지기를 사용하여 지상국의 방향을 탐지할 수 있게 한 것
NDI	Non-destructive Inspection ; 비파괴 검사
NLG	Nose Landing Gear ; 노스 랜딩기어
nm	Nautical Mile ; 해리
NOTAM	Notice to Airman ; 항공 정보, 항공의 여러가지 시설, 업무, 방식 또는 위험 등의 설정 상황 조건의 변경에 관한 것으로 운항 관계자에 대해 지체없이 통보되는 고지를 말함.
NTSB	National Transportation Safety Board ; 미국운수안전위원회

— O —

OAG	Official Airlines Guide ; 전세계의 국내, 국제선의 시각표를 중심으로 운임, 통화 환산표 등 여행에 필요한 자료가 실린 간행물을 말함. 이것은 미국의 출판물인데 같은

내용으로 영국의 출판물인 ABC World Airways Guide 가 있다.

OAT	Outside Air Temperature ; 외기 온도
OC	On Condition ; 상태에 따라

— P —

PA	Passenger Address ; 기내 방송
PAR	Precision Approach Radar ; 정밀 진입 레이더를 말하며 계기 기상 상태에서 최종 진입하는 항공기의 진입로, 강하로에서 벗어남, 접지점까지의 거리를 탐지하여 안전히 착륙시키도록 유도하기 위한 것이다.
P/N	Part Number ; 부품 번호
PSI	Pressure per Square Inch ; 1인치 사방으로 가해지는 1파운드의 압력이나 하중을 말하며 압력의 단위이다.

— Q —

QC	Quality Control ; 품질 관리
QC	Quick Change ; 여객기에서 화물기(또는 이 반대)로 빨리 전환할 수 있는 형의 항공기를 말한다.
QEC	Quick Engine Change

— R —

RCAG	Remote Center Air-Ground Communication ; 원격 대공 통신 시설로서 이것에 의해 원격지의 항공기와 관제 기관의 직접 교신이 가능해진다.

RDI	Radio Direction Indicator ; 콤파스 지시기와 거리 지시기를 합친 계기로 항공기의 자방위, 무선 항로로부터의 편위, 항로의 설정 등을 하는 것이다.
REI	Radio, Electrical and Instrument ; 무선, 전기 및 계기
RH	Right Hand ; 우측
R I I	Required Inspection Item ; 검사 요목
rpm	Revolution Per Minute ; 분당 회전수
RRB	Radio Regulatory Bureau ; 전파 감리국(체신부)
RTK	Revenue Ton Kilo ; 유상 톤 킬로미터. 각 비행 구간의 유상의 여객, 화물, 수화물 및 우편의 중량에 그 구간의 거리를 곱한 것의 합계이다.

— S —

SAE	Society of Automotive Engineers ; 미국자동차기술학회
SB	Service Bulletin ; 서비스 블래틴
SELCAL	Selective Call ; 전화의 호출과 같은 방식으로 항공기에는 호출 부호가 붙여져 있어 지상의 통신국이 이 호출 부호를 발사했을 때만 항공기 내의 부저가 울리는 구성으로 된 장치를 말함.
SFC	Specific Fuel Consumption ; 연료 소비율
SHP	Shaft Horse Power ; 축마력
SL or S/L	Sea Level ; 해면 고도
SQ	Squawk ; 고장 또는 불량 상태
SSB	Single Side-Band

SSR	Secondary Surveillance Radar ; 2차 감시 레이더를 말하며 ARSR 또는 ASR과 조합해서 사용하며, 항공기는 이 장치가 발하는 전파를 받아 그에 대해 기상의 ATC 트랜스폰더(항공 교통 관제용 자동 응답 장치)에서 각기로 특유한 전파를 발사하여 지상의 레이더 스코프상에 항공기의 선택, 식별 및 긴급 사태의 발생 등을 표시한다.
SST	Supersonic Transport ; 초음속 수송기
STOL	Short Takeoff and Landing ; 단거리 이·착륙기

— T —

TACAN	Tactical Air Navigation System ; 극초단파 전방향 방위 거리 측정 장치를 말하며, 군용을 목적으로 개발된 것으로 VOR과 DME의 두가지의 기능을 함께 갖고 있다.
TAS	True Air Speed ; 진대기 속도. 교란되지 않는 대기에 상대적인 항공기의 속도를 말한다.
TBO	Time Between Overhaul ; 오버홀 시간 한계
T/O	Take off ; 이륙
TO	Technical Orders ; 정비 기준(미국 등에서 사용되고 있다)
T/R Unit	Transmitter and Receiver Unit ; 송수신기
TSFC	Thrust Specific Fuel Consumption ; 단위 추력당 연료 소비율
TSO	Time Since Overhaul ; 오버홀로부터의 비행(사용) 시간
TT	Total Time ; 사용 개시 후의 총비행(사용) 시간

— U —

UFO	Unidentified Flying Object ; 미확인 비행 물체
UHF	Ultra High Frequency ; 극초단파
URR	Unscheduled Removal Rate ; 부정기 제거율

— V —

V	Voltage ; 전압
VFR	Visual Flight Rule ; 시계 비행 방식
V/G	Vertical Gyro ; 버티칼 자이로
VHF	Very High Frequency ; 초단파
VOR	VHF Omni Range ; 무지향성 VHF 항법 보조 지상 스테이션
VTOL	Vertical Take off and Landing ; 수직 이·착륙기

— W —

W/B	Weight and Balance ; 중심 측정
WECPNL	Weighted Equivalent Continuous Perceived Noise Level ; 가중 등가 평균 소음 레벨
W/R	Weather Radar ; 기상 레이더
Wx	Weather ; 기상, 기후

저자 약력

조용욱　금오공고 졸
　　　　　대한항공 근무
　　　　　미국 Northrop 대학졸
　　　　　교통부 항공 정비면허 소지
　　　　　미국 FAA 항공 정비면허 소지

한석태　금오공고 졸
　　　　　대한항공 근무
　　　　　교통부 항공면허 소지교통부 한정면허(B747.747~400) 소지

항공영어

1993년 3월 5일 초판발행
2022년 2월 28일 재판발행

　　　　　　저 자　조용욱 · 한석태
　　　　　　발행처　청 연
　　　　　　주 소　서울시 금천구 독산동 967번지 2층
　　　　　　등 록　제18-75호
　　　　　　전 화　02)851-8643
　　　　　　팩 스　02)851-8644

정가 : 20,000원

* 저자의 허락없이 무단 전재 및 복제를 금합니다.
* 낙장 및 파본은 바꿔드립니다.